Surfing Spaces

The act of surfing involves highly-skilled humans gliding, sliding, or otherwise riding waves of energy as they pass through water. As this book argues, however, this act of surfing does not exist in isolation. It is defined by the cultures and geographies that synergize with it – by the places, ideas, images, and other representations which at once reflect, create, and commodify this spatial practice.

This book innovatively explores the spaces of surf and surf-riding, informed specifically by the perspective of human geography. Based on a range of critical turns within the social sciences, the book explores the locations, relational sensibilities, and transformative nature of surfing spaces, and examines how the spatial practice has been scripted by dominant surfing cultures. The book details how prescriptive (b)orders of access, entitlement, and marginalization have been created, and how, with the advent of new craft, media, and ideals, they are being actively challenged to redefine surfing spaces in the twenty-first century.

Jon Anderson is a Professor of Human Geography in the School of Geography and Planning, Cardiff University, UK.

Routledge Research in Culture, Space and Identity
Series Editor: Peter Merriman

The *Routledge Research in Culture, Space and Identity Series* offers a forum for original and innovative research within cultural geography and connected fields. Titles within the series are empirically and theoretically informed and explore a range of dynamic and captivating topics. This series provides a forum for cutting edge research and new theoretical perspectives that reflect the wealth of research currently being undertaken. This series is aimed at upper-level undergraduates, research students and academics, appealing to geographers as well as the broader social sciences, arts and humanities.

Artistic Approaches to Cultural Mapping
Activating Imaginaries and Means of Knowing
Edited by Nancy Duxbury, W.F. Garrett-Petts and Alys Longley

Geopoetics in Practice
Edited by Eric Magrane, Linda Russo, Sarah de Leeuw and Craig Santos Perez

Space, Taste and Affect
Atmospheres That Shape the Way We Eat
Edited by Emily Falconer

Geography, Art, Research
Artistic Research in the GeoHumanities
Harriet Hawkins

Creative Engagements with Ecologies of Place
Geopoetics, Deep Mapping and Slow Residencies
Mary Modeen and Iain Biggs

Comics as a Research Practice
Drawing Narrative Geographies Beyond the Frame
Giada Peterle

Spaces of Puppets in Popular Culture
Grotesque Geographies of the Borderscape
Janet Banfield

For more information about this series, please visit: www.routledge.com/Routledge-Research-in-Culture-Space-and-Identity/book-series/CSI

Surfing Spaces

Jon Anderson

LONDON AND NEW YORK

First published 2023
by Routledge
4 Park Square, Milton Park, Abingdon, Oxon OX14 4RN

and by Routledge
605 Third Avenue, New York, NY 10158

Routledge is an imprint of the Taylor & Francis Group, an informa business

© 2023 Jon Anderson

The right of Jon Anderson to be identified as author of this work has been asserted in accordance with sections 77 and 78 of the Copyright, Designs and Patents Act 1988.

All rights reserved. No part of this book may be reprinted or reproduced or utilised in any form or by any electronic, mechanical, or other means, now known or hereafter invented, including photocopying and recording, or in any information storage or retrieval system, without permission in writing from the publishers.

Trademark notice: Product or corporate names may be trademarks or registered trademarks, and are used only for identification and explanation without intent to infringe.

British Library Cataloguing-in-Publication Data
A catalogue record for this book is available from the British Library

Library of Congress Cataloging-in-Publication Data
A catalog record has been requested for this book

ISBN: 978-1-138-84934-1 (hbk)
ISBN: 978-1-032-30176-1 (pbk)
ISBN: 978-1-315-72567-3 (ebk)

DOI: 10.4324/9781315725673

Typeset in Bembo
by Taylor & Francis Books

Contents

List of figures vii
Foreword viii

INTRODUCTION 1

1 Welcome to Surfing Spaces 3
2 Turning towards Surfing Spaces 23

PART I 37

3 Locating Surfing Spaces 39
4 Relating to Surfing Spaces 51
5 The Surf-Riding Cyborg 63
6 The Relational Sensibilities of Surfing Spaces 74
7 The Stoke of Surf-riders 89
8 The Event of Surfing Spaces 102

PART II 111

9 The Heterogeneous Histories of Surfing Spaces and their Cultural Colonisation 113
10 The Imagineering of Surfing Spaces 125
11 The International Influence of the Surfing Spaces Script 139
12 Coding Surfing Spaces: Surf-rider Positioning 150
13 Codes of Surf-Rider Provenance 160

14	Codes of Craft	171
15	Codes of Gender	182
16	Codes of Travel: The "Trans-Local" Surf-Rider	195
17	New Surfing Spaces	206

Bibliography   220
Index   240

Figures

1.1	"I hankered after the sea as I'd never done anything else before. … being denied access to the ocean was intolerable" (Winton, 2008: 19). (Photo: Tiron Jenkins, eilirhaf.com)	4
1.2	The Illusion of Surfing (Author's own photo).	8
10.1	Composing Surfing Spaces.	130
12.1	The Surfers' Code (Images from Author, and Lyndsey Stoodley).	153
12.2	Coding Surfing Positioning (Image from *Surfer Today*, no date).	154
17.1	Artificial Waves in Wales (Images: Lyndsey Stoodley).	216

Foreword

> As a surfer I understand surfing. But that understanding is deeper, finer, and better contoured for the efforts of Mark Twain, William Finnegan, Susan Orlean, and the rest of the 24-karat-gold writers, nonsurfers as well as surfers, who've examined the sport. And if I do at times feel as if I'm a member of a kingly species, it's partly because I surf, and partly because Jack London said so.
> (Warshaw, 2004: XXI)

Welcome to surfing spaces. Surfing spaces are worlds of practice which involve highly skilled humans riding waves of energy as they pass through water. Due to this emphasis on action, we may be tempted to consider surfing spaces as solely understood through the physical act of surf-riding. However, this book argues that the act of surfing does not exist in isolation, it is defined in part by the cultures that synergise with it — by the ideas, myths, writings, images, and other representations which at once reflect, influence, create, and commodify the practice. As a result, this book suggests that the relations between the practice and the culture, the moment and the reflection, the affect and its subsequent commodification, are mutually definitive. As Warshaw describes above, surfing can be described as a 'sport', but also a 'lifestyle', an 'art', or a 'hobby', and this definition is not incidental or without consequence. That Warshaw refers to surf-riders as 'kings', and the writers he cites are mostly men, is also not a value-free or inconsequential statement either. In the same way, his implication that the voices of surf-riding are predominantly white and American, directly connects the practice to particular geographies and ideas of space, as well as power and (subjugating and colonising) histories. Thus, at once surfing is simply a practice: a human being riding energies that pass through water, but it is also imbricated into a world of power relations, inequalities, conflicts, and geographies. This book explores these practices and cultural relations: welcome to surfing spaces.

Introduction

1 Welcome to Surfing Spaces

Introduction

> Early on, my relationship was formed with the sea. I continue to remind myself of this, wondering where it began, wondering about the origins of my relationship. In attempting to fulfil this search, my mind wanders back to the moments that still remain fresh, moments that are symbolic of my journey. For only in searching back to my origins with the sea will I uncover the reason for my continued exploration.
> (Allen, 2007: 85).

Littoral logbook

Ever since I was a child, and my mother took the family on Sunday drives for what seemed like hours to the coast, it had me. In my naivety, it seemed a long way to go just to get open sky and ice cream. But the journey, the arrival, and the presence of the waves came to matter more to me than the satiation of a sweet tooth. Since then, I have been drawn to the edge of things. I have been hungry for places where the sea, land, and air meet. I love these places, I love this... space. I agree with Camus when he said, "the heart of royal happiness' can be found here (2013: 10). I don't know if my Mum had read Melville[1], but now I'd wager she fully recognised his urge.

Indeed, "a damp, drizzly November in my soul" develops whenever I imagine a world where myself and others are barred from the waves; where we are only allowed to stand on the beach, just this side of the swash and tideline, just this side of the surf. Imagine the exile.

Now, barefoot, I stand here. Watching, then walking, to the water's edge. But this is not just the water's edge. It is the edge of exposed land, the end of (one sort of) culture, and a frontier to a new world. My feet stand, sinking slightly. The sand moves, becomes suspended, stalls, and falls across my toes. My calves are now immersed, with strands of kelp swaying; still; gently brushing my skin. I face the curved horizon, with an exultant off-shore wind lifting my hair and whooshing my ears. Here, at the end of the land, the beginning of the sea, and the cusp of the air, I am at the join, the merger, the specific contact zone where these disparate elements meet and wrestle each other. The intensity of union is such that the modern distinction between these elements simply doesn't matter anymore.

To paraphrase Whitman (2003), this feels like a place of "original energy', and I am "mad" for immersion within it. I take one step, two; the water laps my midriff. For now

DOI: 10.4324/9781315725673-1

4 *Welcome to Surfing Spaces*

Figure 1.1 "I hankered after the sea as I'd never done anything else before.... being denied access to the ocean was intolerable" (Winton, 2008: 19). (Photo: Tiron Jenkins, eilirhaf.com)

at least, I fight every fibre that tempts me to duck dive, float and drift; to give myself up to the currents and tides. Still standing, I watch the shadows on the sea. My heart lifts. The waves are coming. Walls of water, born from the winds out there, memories of sun-driven squalls long died. The sand beneath my feet now mingles round rocks. It is these rapid shallows, this disappearing depth, which draws in the rising dark. The waves rise, form ramparts, yards from me.

I dive into their roots. Into the who knows where, the here and there, the then and now. Surface and depth, far and near, past and future, it all flows round and through me, connecting together in this momentary place. I twist, I turn. Combined energy lifts me to its surface; its last visible act takes me home. Then it is gone. This is the littoral territory, the littoral scape; it is the original space of surf and surf-riding.

The cultural geography of surfing spaces

This book explores the space of surf and surf-riding. As this chapter will introduce, surfing spaces have grown in popularity over recent decades, with growing numbers of participants being involved in all facets of surf culture: from the act of surfing itself, to travel and tourism in surfing areas, to the consumption of surf technologies, and the purchase of related products in the worlds of fashions, media, music, and iconography. As a consequence, surfing has become a major income generator for local surf breaks, whilst the culture surrounding surfing has developed and matured, with surfers taking on key roles in campaigning against pollution, protecting local beaches, as well as selling the "lifestyle" of surfing through corporations to the masses. There are many (non)academic books exploring these myriad facets of surf culture, but this book offers a unique and necessary approach to understand these processes and relations, informed specifically from the perspective of cultural geography.

Adopting a cultural geography perspective to the practice of surfing offers a new and vital approach from which to understand the phenomenon. For many years, geography has ceased to be simply concerned with facts about capital cities, colours of flags, or sizes of population. Human geography, and in particular the sub-discipline of cultural geography, is more concerned with the role of space and place in forming human societies, group identities, and political movements. As I have stated elsewhere, it has become a cliché to suggest that humans are social in nature, but geographers would suggest that humans are also spatial in nature too (Anderson, 2021). If we re-read the littoral logbook that opens this chapter, we can acknowledge how the location of things influenced my formative years – from the journey to the coast, to the meeting of land, sea, and air, the way the resulting waves made me feel, and the role the space in aggregate had to play in my identity formation (even its role in driving me to write this book). In short, we can begin to recognise the difference that place makes in creating who we are and what we do.

Spaces and places therefore play a crucial significance in our lives, as Edward Said notes, "none of us is outside or beyond geography" (1993: 7). But beyond this general acknowledgement of the importance of geography to (y)our identity, more critical questions centre on the way we are connected with space and place, the role geographies have in forming us, and the role we have in forming them. From this perspective, geographies have an influence on the cultural actions occurring in our world, in Tilley's words, place is, "something that is involved in the action and cannot be divorced from it" (1994: 10). This is not to say that geographies somehow *determine* cultural behaviour, but the legacies, values, and contemporary "traces" of cultural actions in particular places often constrain and enable the varieties of cultural life and individual identity that are possible in any given location (see Anderson, 2021). Over time, places are taken and made by different cultural groups, cultures literally take place; they aggressively, passively, intentionally, or otherwise, colonise a place and contributed to its, and our, meaning and identity.

Cultural geography is thus a vital perspective as it reminds us that places are not only a medium but also an outcome of action, producing and being produced through human practice. This perspective highlights how geographies are not simply points on a compass or co-ordinates on a map; they are also cultural – and personal. Geography plays a key role in making our lives meaningful and giving us a sense of identity; rather than being, "highly abstract and remote from experience" (Tuan, 1975: 151) the places around us are "thoroughly meshed" (Casey, 2001: 684) into the human condition. The spaces and places of our lives are, in the words of Preston, "deeply woven into the fabric of who we are" (2003: XVI).

Thus, when we introduce ourselves to surfing spaces, we are, from the perspective of cultural geography, setting out to explore the relations between people, place, and culture that co-define this world. From this perspective, surfing spaces are not simply the practices and the product of surf-riding, but they are always tied into a myriad of other processes: the framing of this practice, the myths and the meanings that are invented and perpetuated through these framings, the politics and power relations that they serve, the entrepreneurship that commodifies them for profit, and the resultant embodied practices that are restricted or mobilised in turn. From this perspective we acknowledge an *interdependent* epistemology where each aspect of surfing spaces is always acting and being acted on by the others. Surfing spaces are thus intensely physical and practice-oriented but also inescapably cultural, each aspect constitutively defining the whole. It is through the coming together of these cultural and geographical phenomena that surfing spaces are composed.

> "[Witnessing surfing for the first time is like] falling in love with a foreign language without being able to speak it or understand it"
>
> (Martin, in Capp, 2004: 218).

Surfing spaces are, therefore, constituted by both practice and culture, and many people are attracted to this "ongoing composition" (after Anderson, 2021). Yet, as Martin outlines above, not all those who are attracted to these spaces are able to fully understand the "language" of these cultures and geographies. Many of us are drawn to the water's edge, to the motion of the waves, and the open horizon. Whilst some are happy to stand and watch, others are compelled to venture into the water and immerse themselves. The aesthetic appeal, the vast openness, and the wild sublime apparently incarnate in this fluid margin, temps our physical engagement. Some are lured into a specific encounter with this littoral; wishing, however fleetingly, to ride the energies passing through the air and ocean, gliding and racing on the surface of the sea. The pull of surfing spaces and the cultural kudos that surrounds them are thus magnetic for many; answering Camus, it seems that large sections of society aspire to be like "these obstinate mad" humans who cling "to planks and [are] tossed by the mane of immense oceans" (2013: 6). Indeed, increasing numbers seek to engage with surfing spaces year on year. As The Economist (2012) and Surfer Today (2013) report, all available figures point towards a growth in global participation in surfing spaces, from 26 million in 2001, to 35 million in 2011 with surfing now practised in over 160 countries across the globe (see Ponting and O" Brien, 2013). The annual surfing industry revenue is estimated to be approximately US$7.29 billion (International Surfing Association 2014). As Barilotti suggests, it seems that today, "everyone wants to be [or be like] a surfer"[2] (no date).

The practices and cultures of surfing spaces thus appear to successfully capture something powerful in the human imaginary. The surf-rider, perhaps once as marginal to mainstream culture as the train spotter or pot-holer, has now become a recognised cypher in the western world. As Mansfield states, "in less than half a century, the surfer has evolved from a creature of social obscurity into a universal icon" (2009: 173). Yet why are so many drawn to this "foreign language", despite being "unable to speak" or "understand" it? By way of introduction, this chapter continues by briefly outlining some of the key practical vocabularies and cultural grammar of surfing spaces.

The language of surfing spaces: Surf-riding

Surf-riding itself involves the successful practice of moving down the surface of a disturbance of energy, materialised as a breaking wave. In its simplest form, surf-riding is the practice of navigating effectively down the face of the wave, not dissimilar to the act of a child whizzing down a slide, or not-quite-grown-up-yet-adult tobogganing on a sled down a snow-covered hill. However, the moment of surf-riding has many variations, dependent by turns on the nature of the wave, the skill of the rider, or the technology employed to navigate it. Where a surfable wave differs from a children's slide or snow-covered hill is its impermanence. In coastal waters, a surfable wave is not fixed in terms of its location, its size, or its structure. Even its existence – when and where it may occur – cannot be guaranteed. The nature of a surfable wave therefore cannot be predicted, but through skill and experience, its location, size, and structure can be anticipated, and responded to. As a consequence, a wave-as-slide could, by turns, be transformed into a curving flume, a racetrack straight, a rutted hillside, or a watery cavern. It could rise in size from a knee-high cascade into a multi-storey collapsing wall. Due to ability or mood, the surf-rider may attempt to simply go arrow straight in front of the breaking wave, or seek to turn along it, gliding or rushing across its face. They may carve up to its breaking crest in order to be in prime position for elaborate aerial or marine manoeuvres; they may dodge emerging bumps and holes, charge hard for speed, or simply seek to survive their encounter.

> The essence of membership of the surfing tribe was each individual's desire to ride waves in the spirit of fun. This was the magic glue that bound us all together as if a tribe. It was exhilarating, intoxicating and quite addictive. It still is
>
> (Mansfield, 2009: 10).

At its heart, surfing is a mobile engagement with breaking water that generates affect in the eye and body of the practitioner. As we will see in Chapters Six and Seven, many labels are given to this embodied affect, but a common description is one of "fun". As Orams and Towner (2013: 174) summarise, "surf-riding is the recreational use of waves for the purpose of propelling the participant forward and attempting to ride these waves *for enjoyment*" (my emphasis). An oft-repeated description about surf-riding, originally attributed to American surfer Phil Edwards (a key figure in the commodification of surfing in the 1960s, see Chapter Ten), is that the best surf-rider is not the one with the most experience, the latest avant-garde equipment, or the flashiest manoeuvres, but "the one having the most fun" (Surfer Today, no date). Even from this apparently neutral, and perhaps even naïve, affective description, a politics of control is identifiable. As we will see in this book (Chapter Nine), this emphasis on surfing-for-fun can be harnessed to actively silence and render obsolete alternative motivations for surfing (and the surf-for-livelihood cultures associated with them, for example those maritime communities who navigate breaks when returning from fishing expeditions), whilst also being flexed to actively include other professional surf cultures (for example highly paid,

8 *Welcome to Surfing Spaces*

lucratively sponsored athletes who compete on world tours, and endorse everything from wetsuits to surfboards, sunglasses to clothing, cars, sports drinks, and watches (etc.)). Thus, whilst the vocabularies of fun retain the kernel of affective truth in defining surfing spaces, it must also be acknowledged that they can also be employed to mould and mask the cultural orders and geographical borders of belonging in contemporary surfing spaces (and for more, see below).

> "Mickey Mouse is a surfer too – on top of a spouting wave at Disneyland, Hong Kong. And so, the illusion of surfing is stronger than ever"
>
> (Hening, 2015: 241).

> "the effect has been made to make the high-risk experience into something akin to a visit to Disneyland or the carnival. There's a big difference. At Disneyland everything is safe. Not so in the outdoors"
>
> (Watters, 2003: 259).

> "It didn't take him long to discover that waves which looked small from the pier got much bigger when you were looking at them from sea level."
>
> (Nunn, 1998: 15).

Although a casual observer may be tempted to perceive surfing spaces as simply and solely places of fun, surfing spaces are not Disneyland. As Rinehart and Sydnor note, surf-riding is "dangerous… it presupposes that bodily risks, including those of fatal accidents, are omnipresent for the participants" (2003: 267). This book argues that these risks are an inevitable consequence of the various elements which combine together to create surfing spaces: the water, land, and atmosphere. Firstly, surf-riders are at risk from the solar energy passing through the water itself – materialised as (rip) tides, currents, and (as Nunn suggests above) wave size. Surf-riders would also be at risk from the presence or absence of solar energy whilst in the water, from the threat of sun burn or heat

Figure 1.2 The Illusion of Surfing (Author's own photo).

stroke, or chill from the cold. Secondly, surfers are at risk from living (but non-human) entities in the water – including sharks, jelly fish, seaweed, reefs, or bacteria (see, for example, Zhou, 2020); thirdly, from other humans in the water – including fellow surf-riders, swimmers, or other water users; fourthly, from other non-living things in the water – including driftwood, debris, abandoned surfboards, or toxins; and finally, from the land itself – including submerged rocks, jetties, and the sea-bed. Thus, as Thorne states, surf-riding,

> can be a very dangerous pursuit, even on relatively small waves, [especially] if the surfer does not have sufficient skill or experience.... Th[is] challenge [is] significant and should be emphasized, as [it is] either entirely unperceived or underestimated by the casual observer
>
> (1976: 209).

However, the dangers inherent in surfing spaces are, for many participants, intrinsic to their cultural appeal. The search for hedonistic thrill (see Midol 1993; Tomlinson 2001; Rinehart and Sydnor 2003; Wheaton 2004) has thus been coupled to the propensity for adrenalin-fuelled risk (see Stranger 1999, 2011) to become an apt description and lucrative commoditisation of surfing over the last 60 years. As surf writer Susan Casey sums up, surf-riding "require[s] a hard squirt of adrenaline – or it [i]sn't really fun" (2011: 11). However, as stated above, the risks involved in surfing spaces are dependent to some degree on the skill of the rider. One fundamental way in which surf-riders heighten their skill levels is through developing a vernacular knowledge of waves' location, size, and structure. This knowledge is hard-earned, requiring dedication, patience, and the attendance to a surfing space and its waves in order to understand – or in surf vernacular become "wired" – about its potential variations and conditions. As we will see, articulating this cultural knowledge, or "oceanic feeling" (after Capp, 2004), is often difficult, as surf-rider and journalist Finnegan explains:

> Nearly all of what happens in the water is ineffable – [normal] language is no help. Wave judgment is fundamental, but how to unpack it? You're sitting in a trough between waves, and you can't see past the approaching swell, which will not become a wave you can catch. You start paddling upcoast and seaward. Why? If the moment were frozen, you could explain that, by your reckoning, there's a fifty-fifty chance that the next wave will have a good takeoff spot about ten yards over and a little farther out from where you are now. This calculation is based on: your last two or three glimpses of the swells outside, each glimpse caught from the crest of a previous swell; the hundred-plus waves you have seen break in the past hour and a half; your cumulative experience of three or four hundred sessions at this spot, including fifteen or twenty days that were much like this one in terms of swell size, swell direction, wind speed, wind direction, tide, season, and sandbar configuration; the way the water seems to be moving across the bottom; the surface texture and the water color; and,

beneath these elements, innumerable subcortical perceptions too subtle and fleeting to express.... Of course, the moment can't be frozen. And the decision whether to sprint-paddle against the current, following your hunch, or to stop and drift, gambling that the next wave will defy the odds and simply come to you, has to be made in an instant

(2015: 76).

Generating this cultural knowledge about the waves' moods and variations is vital to enable the surf-rider to actually catch a wave. Where and how to begin a ride, and how to time one's engagement with the swell, is not a straightforward exercise; as Mark Twain narrates from his first-hand experience: "I got the board placed right, and at the right moment, too; but I missed the connection myself – The board struck the shore in three quarters of a second, without any cargo" (cited in Duane, 1996: 20). The need to locate, then time the "take off" correctly, along with the skill required to ride the board on the swell, makes the practice of surf-riding an expert one. As Fuz Bleakley, a Cornish surfer, confirms, "so many people come to surfing now thinking that they, too, will be deep inside a Pipeline barrel [see Chapter Eight] within just a couple of weeks. They've no idea how difficult surfing is..." (in Wade, 2007: 65).

As a consequence, surfing spaces involve a practice that anyone can undertake, but to be accepted as a surf-rider by those who police the cultural orders and geographical borders of these spaces, one must possess, "sufficient skill and knowledge to utilise the power of a wave for forward momentum, track at an angle across the face of a wave, and anticipate and respond to its changing contours" (Ponting, 2008: 23 cited in Orams and Towner, 2013: 174). Thus to be credibly called a surfer, by yourself and others from the "surfing tribe" (see Mansfield, 2009), it isn't enough to simply be taken for a ride by a wave (even if this can be great fun), but you have to demonstrate the requisite skill and experience to *actively ride* the surface swell, not only determining your direction, but also your line and "style" on the wave (see Chapters Five and Fifteen). As *Carve* magazine's Roger Sharp states, a surfer is someone (but anyone? see below) who can "stand up long enough for it not to be considered a fluke" (2015: 11).

The language of surfing: The technologies of wave riding

We have seen thus far how surfing can be an activity that is undertaken simply for fun but is also a practice dependent on some level of skill and knowledge of surfing space. However, this vernacular knowledge of waves and riding does not exist in isolation; it is necessarily combined with a practical understanding of the most appropriate technologies suited to both rider skill and surf spot conditions. From a layperson's perspective, one might expect that all surfers use a board to ride the waves. Dictionary definitions support this view, suggesting that surfing is constituted as, "the sport of riding towards the shore on the crest of a wave by *standing or lying on a surfboard*" (Oxford Dictionary, 2011, my emphasis). Here we can see that surf-riding is understood to be dependent on a board, regardless of whether individuals choose to stand, kneel, or lie upon it. The Collins English Dictionary concurs with this definition, stating that in order to surf, an individual has to "ride a surfboard" (2003). Academic scholars reinforce this perspective: for Shields, "surfing is the art of standing and riding on a board propelled by breaking waves" (2004: 45); whilst Warren and Gibson define surf-riding as

> "an ancient interaction between humans and the environment, a fluid and exciting pastime where breaking waves, the body, and a *surfboard* interact"
>
> (2014:1, my emphasis).

If these perspectives are taken at face value, one may consider that a board is the only, and perhaps best, way for someone to properly become part of surfing spaces. However, there are many technologies upon which the surface swell can be ridden. Following the definitions above, surfers can be categorised in terms of their *bodily position* in relation to the wave, and also in terms of their *riding technology* (in Chapter Fourteen we will call these categories codes of craft). For example, there can be riders who like to lie to surf (bodysurfers and bodyboarders), those who prefer to sit (or sometimes kneel, including surf-skiers, sit-on-top kayakers or surf-kayakers) and those who stand in order to catch their waves (surfboarders, on either long-, short-, or stand-up paddle-boards). To conflate all surf-riders and surfing practice into the act of board-riding thus, by intention or otherwise, has the effect of ignoring and silencing competing visions about the practice. Despite this, much surf writing, media reportage, corporate advertising, and academic analysis represents the surfing world as one dominated by board-riders (and as we will explore further in Part II, male, white, stand-up, and short-board riders at that). Thus in order to fully comprehend surfing spaces, there needs to be an acknowledgement of the diversity of surf-riders; as Farmer and Short confirm, surf-riding needs to be recognised as including all,

> "bodysurfing, bodyboarding, surfboarding, surfskiing, surfboating, [and] forms of surf lifesaving and lifeguarding, but exclud[ing] all surf interaction powered by wind and machines"
>
> (2007: 100).

As a consequence, this book adopts an inclusive definition of surfing spaces, using the term surf-riding to include any technology predominantly powered by energy passing through the sea.

The language of surfing: Patience and practice

> "Watching the surfers out in the bay, I realised that they do a lot more waiting around in the water than I expected"
>
> (Pretor-Pinney, 2010: 288).

Unlike the rides at theme parks, the waves that pass through coastal surfing spaces are not predictable and do not happen to order. Even in hallowed surf spots such as Hawai'i, "conditions aren't always ideal" (Taylor, 2005: 119). So as Pretor-Pinney states, surf-riders "do a lot more waiting" for swell than actually riding it. Surfing spaces can therefore be more related to waiting for waves, talking about waves, recalling past rides, and dreaming those of the future, than being solely focused on the act of surf-riding. This combination of patience and practice has a number of consequences for surfing spaces. Firstly, the necessity of patience – for swells, waves, and opportunities to ride them – leads to a cultural value being accorded to surf spots that break in specific swell, tidal, or atmospheric conditions. It is not simply the regularity of rideable waves that is lauded, but often the rarity of them; kudos is earned by those who wait for them, then catch them. As surfer Mark Harris describes in relation to British surf culture, "what makes th[is] scene unique is just waiting for the swells, definitely. It makes it even more special when you score in one of the spots [that seldom break]" (Croker and Dean, 2012: 1.50). As a consequence, (and as we will explore in more detail in Chapter Six), when the waiting is over, rideable waves become hard won moments to celebrate; for many, waves are not simply a predictable, everyday occurrence, but a grail to be sought (see also Chapter Seven) and a treasure to discover.

The necessity of patience for swell and wave also generates a strong sense of commitment to, and connection with, a local surfing space. Over time and with experience, surf-riders get a feel for how weather and swell will result in opportunities for wave riding, and a strong connection is developed between an individual, and the geographical processes and places which enable their practice (see Chapters Four and Five). This sense of spatial identification with a local surf spot may generate a sense of loyalty towards and protection of this area, most obvious when its cultures and capacities are threatened by other potential coastal interests or even non-local surfers (see Chapters Thirteen and Sixteen).

When the sea is flat, aspiring surfers will also turn to online weather maps and surf forecast sites to detect when and where the next swell will occur. Individuals may travel along the local coastline and deploy their vernacular knowledge to discern the nature of the rocks, sandbars, and prevailing tides and weather in an attempt to second guess where waves may break in upcoming conditions. When their patience runs out, surfers may even respond to the ephemerality of waves by becoming active explorers; instead of waiting for waves to come to them, surf-riders go to where the waves are breaking. They will plan (un)likely trips to far-flung surf spots, await surf updates from social media, and pore over surf magazines, in lieu of the encounters that mean the most to them. In these circumstances, surfers change from simply being wave-riders to becoming adventurers who are always on a quest for the next wave.

This waiting and searching, punctuated by rare moments of discovery, brings wanderlust to littoral life. Indeed, love for surfing spaces becomes strongly associated, especially in the dominant media and commercial representation of surfing, with a love of travel to surf. As Chapters Ten and Sixteen discuss in detail, this enjoyment of both mobility on a wave and mobility to a wave have become co-ingredient characteristics of dominant surfing spaces. Since the seminal surf movie *The Endless Summer* (Brown, 1966), which told a story of surfers circumnavigating the globe in search of waves, surfing spaces have come to be as much about the quest for the next wave, as the wave itself; as Chris Mauro (when Editor of *Surfer* Magazine) states, "the essence of the surfing experience is not rooted in the simple act of riding waves, but in the pursuit of them, and everything that comes with those journeys" (in Taylor, 2005: flysheet), and as surf journalist Chas Smith (2013: 3) puts it, "Surf *and* adventure [have become] my identity. I ha[ve] fallen forever in love *with both of them*" (my emphasis). The importance of travel to surfing practice thus raises inherent contradictions within its culture: just as many surf-riders feel attachment to their local spot, they also feel antagonism to others who wish to share it. Yet these same surf-riders experience wanderlust to surf waves beyond their local, and thus surfing practices of one location meet those of another, and the cultural orders and geographical borders of surfing spaces are developed through these coming togethers of surf-riders in practice. Such moments of conviviality, but more often conflict, have significant cultural, environmental, political, and commercial consequences. The language of surfing spaces therefore involves a set of complex practices, activities that require qualification in terms of place, individual experience and skill, and technological choice. They require awareness of the geographical borders which enable waves to form, and sensitivity to the cultural orders which bestow privilege, entitlement, and access.

The language of surfing: The spirit of surf-riding

To ride surf either at home or abroad involves a test of skill, strategy, and ability, often in the face of personal fear and challenging waves. It is this testing of an individual's limits, developing combined mental, physical and psychological skills, in an environment that is often (un)familiar, beautiful and threatening, that is alluring to the increasing numbers involved in surfing practice. Although surfing is often a sociable activity, with riders travelling to, paddling out, and waiting for waves together, when a set hits the relationship between human and the elements is individualised: it is just the rider and the wave (see also Chapter Ten). As Bill Hamilton, a famous surfboard shaper and rider from Hawai'i (cited in Booth, 2001: 84) states, "the act of going surfing is... an experience that has nothing to do with anything except you and the ocean, period" (or as Australian writer Tim Winton's craggy protagonist states in *Breath*, a story about surfing, states: "'Son', he said. "Eventually there's just you and it'" (Winton, 2008: 85)).

Now solitary in the surfing spot, the relation between the surfer and the sea is captured in microcosm. The wannabe surf-rider has to face the question whether their skills, knowledge, and technique will be good enough to withstand whatever the emerging waves will conjure. Participating in surfing spaces thus focuses attention not only on the relations between one person and one place – the potential surf-rider and the potential surfed wave, but also on the key moment in which these potential futures may become translated into emerging realities. Will the wave rise, will the individual's cultural knowledge anticipate it accurately? Will the muscle memory position the craft correctly, and will the human execute the skill, balance, and bravery necessary for success? Will they just get lucky? All these questions will be answered soon: the place is here, and the time is now. As Winton's protagonist warns, at this moment, "You'll be out there, thinkin, am I gunna die? Am I fit enough for this? Do I know what I'm doin? Am I solid? Or am I just… ordinary?" (Winton, 2008: 86). Surf-riding crystallises this key crucial interrogation into the nature of personal identity in one geographical and temporal "location", and it is the acknowledgment of this test which renders the experience and practice of surf-riding *extra*ordinary. Surf-riders engage with the breaking waves in order to discover and explore their own internal capacities, be they physical or psychological in nature, they surf in order to discover and affirm the potential of their nature. Through establishing person–place relations in the surf zone, riders exert themselves to discover whether they can feel "extraordinary", and successfully inhabit a semi-amphibious, littoral identity.

> "Surfing is the escape into the pure now of it" (Kampion and Brown, 1997: 25).

Surfing therefore involves patience and practical skills, it may involve planning and preparation, but eventually all the equipment, exploration, anticipation, and rhetoric are reduced to a moment of extraordinary encounter with the breaking wave. This intense focus on the temporal and geographical moment of merging with the surfed wave and emergence of the surfer often means that – at that moment – riders forget about their terrestrial lives and find themselves defined anew in littoral space. For many, therefore, surfing spaces offer a portal to a different world, distanced from everyday commitments of job, life, and family. It offers a space defined by forces and philosophies that are alien to the orthodoxy of terrestrial existence that most are used to. As surfer "Noah" in Welsh surfer and writer Tom Anderson's *Acteaon Tide* describes:

> Once I get out behind the breaking waves, it's all release. Shore is just a model now – a different realm, left behind. Things make meaning and take shape as I stroke into a lined up little peeler that stands up to me, its folding crest holding my eyes. Body weight isn't mine anymore, as I rise to the top of a strip of thin, wind-groomed water, edge tilted to the outside rail and drop back in. The sea's pulse moves through me. Confession, absolution. I'm shaking off a burden, rinsing away things that I don't need… . Two more waves, forgiveness in the bag, and I'm into replenishment. Storms thousands of miles away charge me up. Good for anything once a few more of these frequencies have tapped through my life-force. Anything
> (2014: 25).

Thus, for surf-riders, surfing spaces offer an opportunity to escape from the cultural orders and geographical borders of normal life, however fleetingly. Shore life becomes a model, reduced in size by the scale and immediacy of attention demanded by immersion in the littoral, and this new positioning grants the surf-rider a new perspective to understand the mores and ideals of that shore as simply a cultural construction, rather than the only way for life to be. As we have seen, this experience is often described as the chance to recharge, refresh, and reboot their existence, and enable them to continue in their terrestrial lives. In this way breaking waves function as a carnivalesque utopia, a homeostasis mechanism for modernity perhaps – by turning to water, surf-riders are able to return to society more capable of dealing with its pressures and norms. Equally, but perhaps more radically, this carnivalesque capacity of breaking waves offer surf-riders a way of understanding the world that is different from the dominant orthodoxy of terrestrial society; surfing space is a window into a different epistemological perspective. As Duane puts it; "surfing became for me a way of being in the world, a way of seeing not just the shapes and moods of the waves but the very life of this magnificent place" (1996: XIV).

Indeed, referencing the religious has become a key way for many surf-riders to articulate the spirit of their activity (see Chapter Six); as the following excerpts illustrate:

> [In surfing spaces] you go into oblivion.... Nothing matters but you and the board and the wave and this instant in time.... Suddenly all your life is there in this long, long, stretched out wave, you're removed from the past, everything that has been on your mind becomes immaterial, everything goes to jelly, and you feel completely removed from the world around you
>
> (Midget Farrelly, as told to Craig McGregor, and cited in Colborn et al, 2002: no page).

> Sometimes surfing can seem like the most frivolous pursuit in the world. It is, after all, about nothing more than chasing an intensely personal wish for escapism, about spending hour after hour in the water for a few seconds of transcendent experience
>
> (Duggan, 2012: 133).

> "Surfing [has] gripped my soul. At peak moments it was so intense that it bordered the spiritual. It was like a form of worship that upheld the wave, a spiralling pulse of natural energy, as its dominant focus, worthy of praise"
>
> (Mansfield, 2009: 1).

As the quotations above suggest, the practical elements of surf-riding, its waiting, planning, essential knowledge and skill, allied to the ephemeral and unpredictable union of land, sea, and sky is often articulated by surf-riders using quasi-religious descriptions. Echoing the word's use in the virtual realm, surfing has come to be defined as not simply as a quest for a wave in and of itself (although this forms the basis for surfing practice), but "to look for something' or "to search", in this case, for something transcendent. Through looking for various forms of happiness, surf-riders have found their own personal nirvana in the waves.

Through engaging with surfing space, surf-riders begin to see the world from the lens of the littoral (see Chapters Three and Four, and below). Surfers are sensitive to the "intangible admixture of landmass direction and underwater contours and magic and sin and gods and goddesses…" (Smith, 2013: 19) that come together to form the ephemeral experience of surfing spaces. Surfers are therefore perceptive of the provisional and emergent nature of their littoral world; rather than knowing a world that is fixed, stable and predictable, they engage in a reality that is anything but. As Irish surfer Seamus McGoldrick puts it: "You can be as logical as you want but when you talk about the wind and waves and the ocean, these elemental things go beyond that… it is [just] too chaotic to understand" (in Duggan, 2012: 116). Surfing spaces are thus experienced, sensed, and known as a materially and metaphorically fluid world, where reality is experienced differently, and the possibility of a new approach to life is encountered. Surfing spaces thus not only offer an alternative set of metaphors for living; from California's entrepreneurs – "The surfer is out there in space, fully alive, under the lip, on a great ride, by himself [sic] – beautiful and scary – like life" (Severson, 2014: no page), to American surf writers – "…a surf break can be Walden Pond, a material synecdoche of all one finds mysterious and delightful about the world…." (Duane, 1996: 23), surfing spaces can become a vocabulary and a syntax which is used to understand the nature of human existence and our "interconnectedness" with other processes and life forms on the planet (after Capp, 2004: 198).

In this view, surf-riding and the littoral scape have become extended and developed from a simple, perhaps superficial, practice on the surface of the sea into a "metaphor for life". As discussed in Chapter Three, due to its materiality and allied to the marginality with which the practice and its geography is positioned in relation to the mainstream, surfing spaces have become experienced and known as open spaces, fit for new, blue, sea-sky thinking. The "perfect and absolute blank" that once accurately described these spaces in the modern cultural imagination (see Anderson and Peters, 2014), has become a clean sheet of swell on which riders can write new futures. Surf-riders can be understood as "edge riders" between the land, air, and the sea, living in the "in-between" in order to find a space to see things differently, to carve new lines, imagine different worlds, and create novel languages and categories of existence that are beyond Cartesian dualisms. The escape, hope, and freedom incarnate in this space, alongside its beauty and the thrill of the practice itself, has become a potent blend of attraction, aspiration, and accessible freedom.

The language of surfing: The geographies of waves.

As we have seen throughout this chapter so far, this book locates surfing spaces as a broadly "oceanic" activity (see also Kampion and Brown, 1997). As we will see in Chapter Three in more detail, this book suggests we can be more sophisticated in this definition, and locate surfing specifically in the littoral, coastal zone. This broad and then specific location for the geographies of surfing spaces is an important clarification. It is, of course, possible to ride naturally-occurring waves in many other spaces in the hydrological cycle. It is increasingly popular, for example, to go beyond the littoral and venture offshore to engage in deep-sea surfing. (This practice often accesses larger waves and employs teams of riders on petrol-powered wave-skis to "tow-in" board riders to wave crests.) Beyond oceanic space, surf-riding can also occur on naturally-occurring estuarine bores, channelled river waves, and increasingly in artificial wave parks. The proliferation of wave parks and the nature of their cultures and geographies illustrate how these alternative surfing spaces challenge many aspects of surf-riding as it will be introduced in this book. These artificial surfing spaces change the geographical location of surfing (from the coast to inland, and potentially from outdoors to indoors); they alter the direct connections between waves and natural cycles (e.g. the generation of waves through solar heating, pressure change, wind generation, tidal and current influence, and associated erosion and deposition of rocks and sand on continental shelves (see Chapter Three)), and they remove the skills necessary to identify where and when waves will break (as "artificial" waves can now be designed, created, and timed to order). In short, wave parks usher in a new generation of surfing spaces and raise the need to become more sophisticated in the categorisation of the geography of this activity. (Indeed, it may be that littoral and oceanic surfing soon becomes known as "wild" surfing, complementing the nomenclature of swimming in coastal, riverine, and ocean spaces (on the latter see Bottley, 2019) and thus categorically distinguishing it from surfing in wave parks). It is suggested, therefore, that the advent of these "new" surfing spaces will fundamentally change humans' relations with waves, and the cultures associated with them. In focusing on the "language" of more "conventional" surfing spaces, this book marks a point where the coastal zone remains the dominant and most prominent of surfing spaces, with surfing cultures not yet being wholly influenced by the identification, invention, and diversification of "new" surfing spaces. Indeed, it is the success of the traditional cultures and geographies of surfing that drives diversification into alternatives, and thus it is important to draw a line in the sand and determine our contemporary position in the development of surfing spaces, at the moment where their cultures and geographies are changing again.

The structure of Surfing Spaces

In this opening chapter, we have seen how the language of surfing spaces has come to initially inspire and go on to influence many people. We have seen how both the physical spaces of surf-riding, and the human practices in those spaces, affect the identities of those who participate in the littoral zone; in the language of cultural geography, we have seen how the littoral and surf-riding come to create a constitutive co-ingredience of "human-and-surf-space". We have also glimpsed how the emergence of surfing spaces is not a natural process, but a cultural one. In the language of the sub-discipline, it is a process of cultural ordering and geographical bordering, where one cultural group's ideas of "good" and "bad" or "right" and "wrong" are transformed through practice and prescription from being merely one alternative, into the only way to be. Through this process, cultural ideas are realised in physical form as they take and make the geographies of surf-riding. Surfing spaces are therefore, from the perspective of this book, both the process and the outcome of contestations between different ideas of what surf-riding is and should be; in some cases this may be a convivial coming together of diverse ideas and values, in others an ongoing conflict where one group aims for absolute domination. Surfing spaces therefore do not emerge from a vacuum, rather their "gatherings, weaving, and assembling are all subject to, and productive of, the influence of power" (Cresswell, 2019: 197). This book critically examines these "gatherings, weavings, and assembling" in more detail, unpacking the power that creates the cultures and geographies of surfing spaces.

It does so through the following structure. Building on Chapter One, Chapter Two critically introduces the theoretical turns which enable this book's engagement with surfing spaces. In detail it outlines the spatial, relational, affective, and hydro turns which have driven the social sciences in recent years, and demonstrates how the growing acknowledgment of these aspects of the human condition synergistically inform the type of cultural geography approach adopted by this book. Chapter Two goes on to detail the nature of this "relational cultural geography" (after Anderson, 2021) and in particular its utility not only in helping to understand surfing spaces, but also to structure the remainder of the book. Chapter Two concludes with discussions on my own positionality with respect to the power relations which are productive of the cultures and geographies of surfing spaces, and outlines the utility of a "third space" positionality for negotiating the relational cultural geographies at play in these worlds.

From here, the book is split into two Parts. Part I, Assembling Surfing Spaces, critically examines the ways in which physical geographies assemble with human practices in order to create surfing spaces. Chapter Three introduces the littoral zone as the primary space of surf-riding, framing it as the spatial consequence of the meeting of land, sea, and air, as well as of more-than-human energy flows; it suggests that littoral locations should be understood not as straightforward, fixed, or land-oriented locations, but as thoroughly relational and emergent assemblages. The remaining chapters in Part I go on to discuss how the potential for "tremendous complicity" between humans and place emerges through surfing practice, demonstrating how water worlds in general can "thoroughly interpenetrate" many humans' sense of self (after Shields, 1991: 14). Chapter Four focuses directly on how, through practical encounter and involvement, hydro-humans emerge with and through saline spaces, and littoral locations in particular. It demonstrates how littoral locations come to co-constitute the surf-rider, arguing that there is a generative relation not only between land, sea, and air which create spaces to surf, but also between the human and these geographies which create surfing spaces. It outlines how, as a consequence, the generic term "water worlds" in practice become a nuanced, multi-scalar nexus of attachment and co-constitution for surf-riders, with different intensities of connection built up for different bodies of water in different places, through anticipation, experience, and practice. Chapter Five introduces the range of different "ontologies" that are created when different surf-riding technologies are used in the littoral zone. This chapter argues that surf-riding technologies are key prostheses which extend and disperse the capacities of the human body, enabling it to surf-ride in different waves and in different ways. It will argue that these human-technology assemblages can be understood as "cyborg" in nature (following Haraway, 1991), chimeras of muddled multiples including but not limited to what we once considered as narrowly "human". Chapters Six and Seven explore the different ways in which these cyborg co-constitutions are articulated by a range of surf-riders. Including ideas of "flow", "convergence", and "religion" (in Chapter Six), as well as "stoke" – a catch-all term that has risen to prominence in modern surf-riding to describe this range of "relational sensibilities" (in Chapter Seven), these chapters examine the variety of different affects which arise in the surf-rider as a consequence of their engagement with surfing spaces. Finally, Chapter Eight employs Badiou's conception of the event (2005) to summarise the key insights from Part I. It argues that the physical geographies and human practices which are co-constitutive of surfing spaces can best be understood as transformative in nature, inspiring many to orientate their lives in order to regularly encounter and experience the breaking wave.

Thus Part I makes the case that the coming together of human practice and physical geography creates a range of relational sensibilities that make surf-riders feel extraordinary. Yet surfers may also be affected in this way not simply because of their own encounters and experiences, but because of the range of cultural commentators, writers, and media which frame their activity. As Warshaw identifies for us in the Foreword, "if I do at times feel as if I'm a member of a kingly species, it's partly because I surf, and partly because Jack London said so" (2004: XXI). In this light, Part II draws our attention directly to the role of culture in surfing spaces, accounting for the ways in which the transformative encounters which come to define surfing spaces have been (b)ordered by those involved. As we have seen above, and Chapter Two will outline in detail, it is possible for one group within a culture to attempt to establish its own ideas and values – its own language if you like, as the normal and taken for granted lexicon of all within their culture. Part II documents how, in the twentieth century, one particular version of the surfing space assemblage came to prominence, with its own language dominating how surfing spaces are understood globally. *Scripting Surfing Space* details this process. Specifically it argues that one particular version of surfing spaces rose to prominence through an act of "imagineering". This process involved the experience of wave-riding being transformed into a "script" which infused surfing spaces with ideas of freedom, autonomy, and individualism. This script resonated strongly with western archetypes of open frontiers and self-determination, and successfully employed a range of media "architecture" to disseminate its values throughout Western America, and then to key surfing markets around the world. This surfing spaces script in turn created cultural codes of practice which sought to define the specifics of what surfing spaces could and should be (the *who*, the *how*, the *where*, the *why*, as well as *which* technologies were most appropriate within surfing spaces). These codes, their imposition, adoption, and subsequent perpetuation completed the imagineering process as prescriptive (b)orders of access, entitlement, and marginalisation were realised in practice.

However, this book argues that by the second decade of the twenty-first century, the domination of this hegemonic script was weakening. Although the script and its codes remained generally consistent and intoxicating for many, the cultural, political, and media context in which they circulated had significantly changed. As a result, the internal contradictions of the script and its codes had become increasingly difficult to justify and sustain. Part II details this process of scripting, dominance, and contradiction in the following way. Chapter Nine outlines how historically surf-riding was a polyvocal cultural activity – with a range of different surfing languages emanating from the variety of identities, positions, and geographies that are outlined in Part I. The chapter progresses to account for the rise to power of one of these languages of surfing space, by recounting the well-documented colonisation, marginalisation, and appropriation of endogenous Hawai'ian surfing. Chapters Ten and Eleven then document the translocation of this creole surfing language to the United States. Collectively these chapters argue that a new surfing space script occurred due to the selective re-packaging and production of key aspects of surf-riding up to that point. Specifically, Chapter Ten outlines how this process scripted surfing spaces with western values of freedom, autonomy, and individualism, imagineering the surf-rider as a heroic protagonist who, through his brave and dramatic encounters, was able to escape the confines of mainstream culture. This script had the effect of elevating all aspects of the surfing space assemblage from the realities of the contemporary, and appealed strongly to the zeitgeist of both "rebellious" youth cultures of the 1950s, and later the counter-cultural movements of the 1960s. Chapter Eleven then documents the growth in popularity of this script, demonstrating how through the use of media architecture it came to define what we now understand as the cipher of "the surfer", as well as generating the role of "the surf-lifestyle entrepreneur" and the now lucrative and globalised "surf market".

Yet as we will see, the scripting of surfing space as aspirational and escapist, meant it was also simplistic and illusory. This combination of characteristics meant that tensions were created not only within surfing spaces, but also between surfing spaces and broader cultural geographies. In practice, littoral locations were not free and open spaces, rather they were thoroughly "coded" with (b)orders, imposed and perpetuated (in part at least) by the imagineering process. In a similar way, any surfer who valued their connections to broader goals of equity, justice, or sustainability, often felt that the simplistic imaginary of the surfing spaces script did not altogether cohere with their broader politics. With these tensions in mind, Part II continues by documenting the nature of the codes that have come to (b)order surfing spaces. Chapter Twelve initially introduces the concept of cultural "codes" before detailing how, in relation to surfing spaces, they can best be understood around two key orientations: the geographical *positioning* of the surf-rider on the wave, and the cultural *positionality* of the surf-rider. This chapter progresses by critically discussing the importance of geographical *positioning* on the wave and outlines how this positioning valorises a particular version of the surf-rider (specifically the young, white, male), and particular styles of surf-riding (specifically aggressive short-board surfing). The chapters that follow focus on the *positionality* of the surf-rider specifically, and the ways in which particular cultural ideas of socio-demographic eligibility have leached into surf-riding codes.

Chapter Thirteen explores the (b)orders of privileged access that are claimed by surf-riders based on the experience, competency, and co-constituent relations that they have built up in their "local" breaks. It outlines how this sense of local "provenance" is often accompanied with a sense that incoming surf-riders may disrupt its privileges. As a consequence, codes of provenance commonly demonstrate protectionist tactics to deter outsiders and maintain local's dominance. Chapter Fourteen further explores codes of surf-rider positionality, but specifically in terms of technology. It examines how particular craft materialities (i.e. the short-board) and craft styles (i.e. speed) have come to prominence, instilling a prejudicial and marginalising "politics of verticality" into surfing spaces. In turn, Chapter Fifteen explores the code of surf-rider positionality in relation to gender. It examines how the script, codes, and media architecture of surfing spaces have enhanced the idea of the superiority of men (as surf-riders), and in particular a specific type of masculinity (in surf-riding). It identifies how all bodies that do not conform with these dominant idea(l)s have been abstracted from the waves, and are only legitimised by men, on masculine terms. Chapter Sixteen details how the wanderlust that is inherent to all surf-riding was scripted into surfing spaces, and in turn, imagineered into a "trans-local" surf-rider code. It argues that this positionality has the effect of extending and compounding inequitable, unsustainable, and colonial assumptions about destination littorals, in effect perpetuating reductive and racist value systems through the imposition of the "dream" script (see also Laderman, 2014).

However, as stated above, in the twenty-first century the internal contradictions of the surfing spaces script and its dominant codes have increasingly become difficult to justify. Therefore, from Chapter Twelve onwards, we will also document the challenges to the codes of positioning and positionality that are emerging in surfing spaces. Respectively, the chapters will critically examine how some local surfers seek to move beyond considering outsiders as threats to the purity of their local surfing space. We will examine how changes in craft complicate and subvert hierarchies of surf-rider positionality, as well as how the reductive positioning of woman-as-surfing-accessory is being wholly challenged by a range of practices, including the establishment of media architecture that cater specifically for women (and indeed other) surf-riders who seek a new version of the surfing spaces script. We will also explore how some travelling surf-riders explicitly jettison the "trans-local surf-rider" code in order to connect differently to the destination littorals they encounter.

In Chapter Seventeen we connect these challenges to the dominant surfing spaces script and align them with the contemporary context of surf-riding. With the advent of new and changing socio-demographics in surfing, the development of new technologies and craft, the discovery and invention of different physical spaces in which to surf-ride, and new media in which communicate and imagineer it, Chapter Seventeen argues that the codes and script that have dominated surfing spaces since the 1950s have never been more threatened. It is argued that the script, codes, and media architecture of surfing face an existential crisis – and it is uncertain what the future of surfing spaces may be. Stemming from its roots in a range of endogenous histories, and now branching out in a range of new cultural geographies, what will surf-riding in the next generation be like? How can the stoke of surfing be maintained, and how will it be (b)ordered, by whom, and to what ends? The final chapter explores these issues, offering key questions which form a new research agenda for the surfing spaces of the future. Dive in…

Notes

1 It may have become somewhat of a cliché to quote this source, but to me it nevertheless points towards the profound draw of the sea; as Melville puts it:
"I thought I would…see the watery part of the world. It is a way I have of driving off the spleen, and regulating the circulation. Whenever I find myself growing grim about the mouth; whenever it is a damp, drizzly November in my soul… then I account it high time to get to sea as soon as I can. This is my substitute for pistol and ball" (1992: 3).
2 Indeed, although industries associated with selling specific surfing products (for example wetsuits, surfboards, etc) are money-making, it is the broader appeal of the surfing space imaginary where more significant profits are realised. It is not simply active surfers who buy surf-related t-shirts, hoodies, sneakers, or sunglasses, but a range of armchair surfers too. It is this latter market that generates the most money; as Surfer Today notes, "Surf companies don't live off selling [active surf] gear, that's for sure. [For] the 'Big Four' – Billabong, O'Neill, Quiksilver and RipCurl – [spin-out] surf wear is where the profit is. Forget wetsuits, surf fins, traction pads or surfboards. Surf brands want you to buy boardshorts, shoes and sandals, sweatshirts, t-shirts, trousers. That's where the coin meets the pocket" (2014).

2 Turning towards Surfing Spaces

Introduction

As we have seen in Chapter One, this book adopts a cultural geography approach to surfing spaces. Critically investigating the cultures and geographies of surf riding is, as an object of study, relatively new to the social sciences. Perhaps over ten years ago, studying surfing spaces would be seen to be as marginal as the littoral zone itself. Reflecting this perceived marginality, historically there have been "concerns about the legitimacy of researching surf culture" (Hill and Abbott, 2009: 276), and as a consequence, surfing "has not been widely researched and is scarcely reported in the academic literature" (Orams and Towner, 2013: 173). More specifically, despite some notable exceptions, geographers "have paid little attention to surfing" (Waitt, 2008:77). However, what disciplines study and how they study it evolves over time, and, over the last decade, this situation has substantively changed. As we have seen in Chapter One, the growing popularity of surf riding has been one external driving force for the burgeoning academic interest in this area, and this has contributed to a loosely defined arena of "surf studies" (after Hough-Snee and Eastman, 2017), with scholars bringing a range of theoretical faculties to the critical investigation of the economic (from processes of corporate commodification, see Stranger, 2011, to the monetary value of surf breaks, see Lazarow et al, 2009), the political (including processes of local territorialism, see Scheibel, 1995, Preston-Whyte, 2002, Daskalos, 2007), the environmental (including pollution and activism, see Ward, 1996 and Wheaton, 2007), and the social (see, for example, Ford and Brown, 2006 and Evers, 2009) dimensions of surfing spaces. For human geography specifically, these external forces have combined with a range of re-orientations within and beyond the discipline which have stimulated interest in surfing spaces. This chapter briefly introduces these re-orientations – or turns – to demonstrate how they inform the cultural geography approach of this book. Specifically, it outlines the "spatial", "hydro", "embodied", and "relational" turns which have driven the social sciences in recent years, and through doing so, demonstrates how the growing acknowledgment of these aspects of the human condition synergistically inform the type of cultural geography approach adopted by this book. The chapter goes on to detail the nature of this "relational cultural geography" (after Anderson, 2021) and its utility in helping to understand surfing

DOI: 10.4324/9781315725673-2

spaces. The chapter concludes with discussions on my own positionality with respect to the power relations which are productive of the cultures and geographies of surfing spaces, and outlines the utility of a "third space" positionality for negotiating the relational cultural geographies at play in these worlds.

A re-orientation towards geography: The spatial turn

Over recent years, a spatial turn has been evident across the arts, humanities, and social sciences. Due to broader processes such as industrial change, capitalist reorganisation, and the related intensification of globalisation, the importance of place in our everyday life is increasingly acknowledged. As capital spreads across the globe, as jobs and industries relocate, and as people become less anchored to particular locales and become increasingly mobile, the relationships between people and places changes. Globalization through capital, militarisation, and environmental hazard has accentuated the significance of location (after Warf and Arias, 2008) and this process has implications for the geographies of ethical, political, and economic interdependence, as well as for human attachment to place. As Lossau and Lippuner (2004) identify and as Levy states, the

> "relevance of space in human life is increasing....[there is] growing pressure on scientists to produce analyses of and insight into the spatial dimension of society"
>
> (2004: 133).

The relevance of place and space for human life in the era of globalisation has led to a "geographical turn" in the social sciences (see Martin, 1999; Agnew, 1995). As acknowledgement grows that the geographical dimension of our lives "has never been of greater practical and political significance" (Soja, 1996: 1) a range of disciplines across the social sciences and humanities are exploring the ways we think about space and place. As Warf and Arias put it:

> Across the disciplines, the study of space has undergone a profound and sustained transformation. Space, place, mapping, and geographical imaginations have become commonplace topics in a variety of analytical fields. ... While this transformation has led to a renaissance in human geography, it also has manifested itself in the humanities and other social sciences
>
> (2008: sleeve notes).

This geographical turn has been practically illustrated in a number of ways. Firstly, as Pugh notes, it is discernible in terms of the subject focus of different disciplines:

> Many of the main topics of debate in the social sciences and humanities [now] include climate change, the worldwide disaggregation of international production processes, global human rights, liberal democracy

promotion, fair trade, imperialism, postcolonialism [etc.]. New material practices, such as the internet and new media developments, seem only to further turn people's gaze towards the tensions between space as a predefined territorial entity, and as the post-territorial circulation of power relations, contingently articulated political identities and ethical responsibilities

(2009:1–2).

The spatial turn has thus involved geographical subject matters being embraced by a range of academic disciplines, and through this process the nature of disciplinary relations has been reconfigured. The importance of location is now considered not only in a strictly geographical sense, but also in the ways it defines social positionality and identity (Haraway, 1988; Bondi, 1997; Rose 1997). Concomitantly, it has been acknowledged that time can no longer be considered without space (Soja, 1996, Massey, 2006a), the interdependent "trialectics" of space, time, and society having led to the study of key "geographical issues" becoming *inter*disciplinary.

The processes of globalisation and academic reorientation has thus focused attention on the ways in which key geographical terms are conceptualised. It has become crucial to adapt and reformulate notions of space and place to remain appropriate and insightful for the context of the (post)modern world. In short, it has become crucial to make sure we have the best ways to "think geographically" so a range of disciplines, including geography, remain relevant to the twenty first century condition; as Massey points out:

> "The way in which space is conceptualised is of critical significance for the conduct of social sciences… It is that the way in which space is conceptualised is of fundamental importance. It *matters*"
>
> (Massey, 2006: 89).

This (re)positioning of geography as central to the human condition and a concomitant (re)conceptualisation of its key terms of reference has been influenced by a second key turn in the social sciences, the turn to water.

A re-orientation towards the hydro: The turn to water.

As many scholars have formally acknowledged the importance of place to the human condition, others have drawn attention to the array of locations that influence our cultures. Until recently it could be argued that much of human geography has effectively been terrestrial in nature – focusing attention solely on land-based spaces and ignoring aerial or watery worlds. As Lambert et al (2006) sum up, human geography was (until the last decade) very much a "'landlocked' field". In the last ten years, however, scholars have begun to acknowledge and explore the range of non-terrestrial spaces that come to influence contemporary human life, and have not only ventured up into the air (see for example Adey, 2010), but also demonstrated the significance of critically exploring worlds of water (see Steinberg, 2001, 2013; Peters, 2010, 2012).

As Anderson and Peters have argued, there has been an "enduring marginalisation" of water from human geographical inquiry, and this has led to "key lacunae in our understanding of … contemporary and historical life" (2014: 21). Despite the acknowledgement of the key role that all stages of the hydrological cycle have for our everyday existence, an awareness of the reliance that all western nations have on the maritime (for trade and collective security, for example, both historically and in the present day), and the knowledge that many societies not only rely on but also orient their worldview to the water, the dominant assumptions of modern knowledge systems remain terra-centric in nature. This enduring marginalisation contributes to an ongoing normalisation of the complacency with which water worlds are regarded, and disregard for the ongoing precarity of our relations with it.

The marginalisation of these media has consequences for not only what we know about the world, but also how we consider the nature of the world itself. Modern knowledge has not only "locked" onto the "land" as the primary object of inquiry, but also been captivated with an ontological approach which mirrors the apparently fixed and locked materiality of terra firma. According to Cresswell (2006), the dominant way in which the modern world is studied is through adopting the language of a "sedentary metaphysics" (after Malkki, 1992). This approach reflects and reproduces stable ontological conceptualisations of the world, and discourages us from acknowledging, let alone diving into, the fluid materiality of water. Yet as scholars have cast themselves adrift from the limits of such modern metaphysics, they have re-thought their traditional geo-graphy (or "earth writing") in terms of its subject matter, and its approach. Acknowledging for example how water worlds are not arenas of fixed and durable entities, but are primarily defined by fluid, unstable processes (as seas mutate from still calm expanses into waves, tides, and storm surges, but are also changeable in terms of their physical chemical state, perhaps as either solid (ice), liquid, or water vapor), scholars have embraced the water world not simply for its importance for human life, but also for its fluid ontology. As such, this turn to water has begun to overturn the "privileging of stable [ways of thinking as] a central tenet of Western metaphysics" (after Deleuze, in Dovey 2010: 22). In its place, scholars have been taken by ripcurrents of flow, change, and hybridity. As our ontologies mutate with our media, and our metaphysics turn from *geo*-logics to *hydro*-logics – scholars have begun to consider how our world could be more appropriately categorised not as fixed but fluid, with all "things" we once considered to be durable and forever, only temporarily stabilised in their mobility, trajectory, and constitutional state.

A re-orientation towards the body: The turn to embodied practice

The spatial turn has thus challenged the modern separation of humans from the world, whilst the watery turn has challenged the "land-locked" nature of the social sciences, arts, and humanities. These turns have gained further momentum when considered in conjunction with a third academic reorientation that is identifiable over recent decades: the turn to embodied practice. In the turns briefly acknowledged above, we can identify the importance of an apparently "simple" change in vocabulary to how the world is re-presented. To emphasise "immersion" for example, not only challenges us to reconsider our relations to the world of which (we are now) a part, but also reminds us that our bodies are instruments through which we can register, sense, and know the world. Such an apparently straightforward choice of verb can thus be philosophically significant: a word can (re)create a world. Indeed, emphasising the role of the body is part of the destabilising and dissolution of what Latour (1993) describes as the "modern constitution". According to Latour, modernity constitutes the world through a false identification, ontological separation, and subsequent privileging of various components within the world. This constitution has attained the status of orthodoxy, becoming a "noun chunk", fossilised and familiar in the contemporary mainstream (after Laurier and Philo, 1999). Prior to the spatial turn, for example, it was considered normal for "humans" to be identified and differentiated from "the world", and subsequently elevated and privileged as superior in terms of value and meaning in this relation. In a similar vein, the modern constitution separated and privileged the human mind from its corporeal body as a primary and only legitimate source of knowledge. Human knowledge was accepted as originating from the cognitive processes of the rational mind, and that which was sensed by the emotional body was deemed irrelevant and silenced. In short, the human condition as constituted by the modern was not only detached from the world, but also "disembodied" (McDowell, 1994: 241), producing a knowledge that functioned as a faithful re-presentation of detached and distant human-world relations.

However, as we will see, a number of non-, post-, and a-modern scholars have sought to question this modern constitution. These scholars take issue with the isolation of the mind as the best and only route to knowing, and recoil at the concomitant deracination of the human body's potential to inform and enrich our sense of the world. As Lewis challenges, "the body [remains] our most fundamental means of communication and interaction with the world. We make sense of the world by acquiring information through our bodies" (2000: 68), and as Merleau-Ponty states, it is:

> my body [that] is the seat or rather the very actuality of the phenomenon of expression... [I]t is the fabric into which all objects are woven, and it is, at least in relation to the perceived world, the general instrument of my "comprehension"
>
> (1962: 235).

Following Merleau-Ponty, Game (1997) notes how our philosophical praxis should explicitly value experience, and through the valorisation of embodied engagement and encounter, emphasise the importance of attachment (rather than detachment) between the self and the world. As Game puts it, we should:

start... with the assumption that we "participate in things": not at a distance, as we are in the world we would know. Thus philosophical reflection involves experience, which implies that the sensuous and the affective are central to the process of knowing: The very formulation "experiencing the phenomenon" places emphasis on experience as the relation between, which not only disrupts a subject–object opposition but implies that experience is constitutive

(1997: 385).

Such challenging assumptions have been developed in the work of "passionate" sociologists (see Game and Metcalfe, 1996, as well as Crossley 1996, Radley 1995; Watson 1998; Clough 2010; Blackman 2012), as well scholars driving the emotional turn in the social sciences (see Anderson and Smith, 2001; Davidson and Milligan, 2004; Davidson et al., 2005; Nash, 2000; Thrift, 1996, 1997, 1999 and Wood and Smith, 2004), and those prioritising the body as a source of meaning and knowing (see Leder, 1990; Mol, 2003). These challenges have led to a re-embodied framing of the world and re-emphasised the importance of siting humans in relation to that world. From these perspectives, humans are no longer simply detached, isolated minds rationally reflecting on our distance and disconnection from the world, but rather we are spatial beings, with our very attachment to the world actively informing our mutual constitution. Such relational approaches (themselves informed by the geographies of Doel (1999) and Murdoch (2006) or the actor networks of Callon (1986) and Latour (1999)), mark the shift away from the independent conceptual categories of the modern constitution and towards an *inter* dependent epistemology where things are always acting and being acted upon by everything else. In sum, they mark a rejection of a static (phenomenological) ontology of "being-in-the-world" (associated with Heidegger, see Guignon, 1993), and an embrace of more emergent and emerging (post-phenomenological) ontologies of "becoming-in-the-world" (associated with Deleuze, 1985; 1993).

A re-orientation towards co-constitution: The relational turn

Thus the spatial, watery, and embodied turns in the social sciences have significance for our studies of surfing spaces. These challenges to the modern constitution, which can perhaps be understood as a more than modern constitut*ing*, suggests that the world is not fixed and separated into noun chunks, but rather a perpetually emerging set of intermingling verbs. As Braun suggests, "the world does not exist of discrete 'things' that are brought into relation, but of flows and connections within which things are continuously (re)constituted" (2004: 171). As part of this ongoing becoming, we too are being reciprocally composed through our engagement with the world (and for those of us who engaged with the surfed wave, this composition is actively informed by our coming together of water, wind, bedrock, swell, and surfing technology, see Chapters Four and Five). Enabled by the rise of postmodern cultural and social theory (after Oakes, 1997: 509), geographers have disrupted and challenged the stable, coherent, and static approach to geographical sites favoured by sedentary metaphysics. As Jones identifies, they all approach the notion of place in a *relational* way:

> "recent years have witnessed a burgeoning work on 'thinking space relationally'. According to its advocates, relational thinking challenges human geography by insisting on an open-ended, mobile, networked, and actor-centred geographic becoming"
>
> (2009: 5).

Relational thinking marks a shift away from the independent conceptual categories of a sedentary metaphysics and the modern constitution more broadly (see Latour, 1993). No longer do the "noun chunks" of this constitution hold sway. The fixed and essential notions of "physical", "human", "culture", or "economy" and their associated binary framings (e.g. "A – not A") give way to an *inter* dependent epistemology where things are always acting and being acted upon by everything else.

From this perspective, the world, and the places within it, are not a given but something immanent, forever forming, and in progress. From this perspective, therefore, place is far from a static, stable, or fixed entity; as Dovey states, "all places are in a state of continuous change" (2010: 3). Place is no longer reliable, consistent, or necessarily coherent; it is wholly provisional and unstable. Place, at any moment, emerges in time and space from the web of flows and connections meeting at a particular node.

As Phillips (2006) outlines, the notion of the assemblage has gained academic currency as a key way to conceptualise these newly conceived relational places (see also Anderson and McFarlane, 2011). The assemblage, inspired by Deleuze and Guattari (1981), and expanded by Delanda (2006), displays many of the key aspects of an amodern approach: it is relational, provisional, and interdependent in its formation. As Dovey suggests:

> In the most general sense an "assemblage" is a whole "whose properties emerge from the interactions between parts" (Delanda, 2006: 5)....The parts of an assemblage are contingent rather than necessary, they are aggregated... ; as in a "machine" they can be taken out and used in other assemblages
>
> (2010: 16).

An assemblage is, therefore, a component that is formed by the coming together of many other parts. These parts do not come together necessarily by intention or design or have an essential permanence that makes their connection insoluble; rather, their aggregation keeps their coherence as an individual unit intact but, nevertheless, forms a larger whole through their connection with others.

Spatial, watery, embodied, and relational turns have thus refocused the "geographical sensibility" (Anderson and Smith, 2001: 9) and opened up the possibility to not only study surfing spaces, but for this study to relationally inform these ongoing (re)orientations. As we have seen above, a shift towards an assembled metaphysics synergises well with the turn to hydrologics across the social sciences. The water world can be well described as a constant state of becoming, it is a world of immanence and transience. As Steinberg and Peters (2015: 248) have argued, taking the "wet" seriously suggests a world of "immanent (self-organizing) event-spaces dynamically composed of bodies, doings, and sayings" (following Jones, et al, 2007: 265), it is a world that is "continually being reproduced" (Steinberg and Peters, 2015: 252), a world composed through a fluid, relational, ontology.

Taken in combination, these turns resonantly remind us of the starting tenets of this book (see Chapter One). Firstly, they emphasise that we, as apparently discrete humans, are not in fact isolated selves; rather, we are spatial beings, part of the cultural world we are studying rather than apart from it. Secondly, they emphasise that a key media in our co-constitutive world is water, and that (thirdly) our bodies are key registers of knowledge about our relations with this world. More than this, however, these turns (fourthly) emphasise how this world and our positioning within it is not fixed; "we" are changing, "it" is changing, and thus we live in worlds of dynamic interaction and interrelation, of merger and emergence. These turns, in sum, emphasise how our cultures and geographies are connected, and although there are times when "we" do feel separate and distinct from "it", in others we may feel these entities/processes merging, and it is not clear where the person and the place begin and end. In these cases, we may wish to re-employ the modern ontological divisions between this "thing" and "that", whilst also acknowledging that new categories may be necessary to appropriately identify the new constellations we are constituted through.

> "We live in worlds of human-created categories, which we modify when new situations arise that call for new responses"
>
> (Cloke and Johnston, 2005: 2).

> "the relation is the smallest unit of analysis"
>
> (Haraway, 2003: 20).

This new interpretation of the language of the world means that the fixed and essential notions of "physical", "human", "culture", or "economy" and their associated binary framings (e.g. "A – not A") may sometimes give way to an *inter*dependent epistemology where things are always acting and being acted on by everything else. As we have seen above, when we focus on the relations and interactions that continually form the world, we emphasise how no-longer-isolated categories are bound together in "networks" (Callon, 1986; Latour, 1999), "assemblages" (Delanda, 2006), or "relational complexes"

(Rouse, 1996). It is through this coming together of different components that places are generated, as Callon and Law suggest, "it is the relations… that are important. [It is the] relations which perform" (1995: 485). In this light, places are brought into being through the creation and practice; as Carter states "places are brought into being performatively" (2018: 49), and as Sheller and Urry confirm:

> Places [could be] presumed to be relatively fixed, given, and separate from those visiting.…Rather, [following the academic turns noted above, it is now understood that] there is a complex relationality of places and persons connected through performances.… Activities are not separate from places… Indeed, places… depend in part upon what is practised within them…
>
> (2006: 214).

The cultural geography approach of this book therefore suggests that all surfing spaces – which we may be tempted to consider as fixed and durable "things" – are rather brought into being through their use (in other words, what they are depends on how they are put into practice, harnessed, and perpetuated – in other words, how they are "performed"). This approach thus involves analysing and interrogating all the agents, activities, ideas, and contexts that combine together to take and make these places. It suggests that the composition of surfing spaces is *always* up for grabs and open to new performances. (B)orders can always be crossed, and new surfing spaces can always emerge.

In order to diagnose the emergence of surfing (or any) spaces, relational cultural geography foregrounds a particular understanding of the "influence of power" (after Cresswell, 2019). In simple terms, power is the "ability to act", but as all cultural acts have geographical consequences, relational cultural geography suggests that power is spatial. All acts of power are performed in, and are productive of, geography; as Cresswell states:

> "power may be used to affect unwanted change in others, and it may be used collectively to accomplish a shared project. [Whatever the goal, p]ower exists in and through place[s]"
>
> (2019: 198).

Due to the inherent spatiality of power, it is through geography that we can identify who has power, how they are wielding it, and to what ends; as Cresswell suggests (above), power can be polyvocal in nature, it can create conviviality, or conflict. Power has the ability to grant access (to all, or to some), it can render eligible, or O/other-wise, it can valorise, it can marginalise; power is both partial, and pejorative. Taken together, we identify power through its ability to order and border the world, including the world of surfing spaces.

As we will see in Part II of this book, when considering how prominent surfing spaces are (b)ordered power can be understood as "dominating" in nature (see Lukes, 1974). This is power that has taken and made geographies in ways that manipulates, encourages, or enforces people to act in certain ways. This may be through the simple construction of signs that request particular behaviours, appealing to the spirit of individuals of "fair play" or cultural respect. Dominating power may be discernible through the threat of cultural exile or "othering" if failing to conform to conventional codes, or the exercise of physical coercion to ensure conformity to what the dominant codes prescribe as appropriate behaviour. From the perspective of relational cultural geography, dominating power is thus performance which is successful at making people obey. It makes people conform to an other group's vision of what the world should be like, whether it is in their interests or not. Dominating power is thus successful in imposing its ideology on particular places. Ideology, as Ryan outlines, "describes the beliefs, attitudes, and habits of feeling which a society inculcates in order to generate an automatic reproduction of its structuring premises" (cited in Wallace, 2003: 239). In other words, ideology comprises the particular ideals and beliefs of one cultural group and presents them as the "common sense" for all.

Dominating power thus works to instil its own values (and "scripts" as Part II details) on the cultural geographies of, in this case, surfing spaces. As Jordan and Weedon tell us, it is through these acts of definition that we are persuaded into learning, "what is right and wrong, good and bad, normal and abnormal, beautiful and ugly" (1995: 5). These scripts are not "true", "neutral" or "natural", but groups attempt to persuade us that they are. As stated above, however, it is always possible for (b)orders to be crossed, values to be challenged, and new cultural geographies to emerge. This occurs when individuals and groups openly "resist" dominating power, intentionally oppose and transgress (b)orders, and seek to reveal and dissolve orthodoxies. Such acts open up spaces to alternative futures, enabling the possibility of thinking and acting differently.

If we take the implications of this chapter seriously, it also involves locating *ourselves* in the worlds that we study. In doing so we are obliged to critically reflect on our own positionality with respect to the power relations which we are subject to, and that in turn are productive of our cultures and geographies. It is to this issue in regard to surfing spaces that this chapter now turns.

Re-orientating positionality in a world that is turning.

As the recent academic turns remind us, the identity of humans (as surfers or otherwise) is never fixed or unchanging. We therefore have to exercise caution in claiming definitive and exclusive positioning for ourselves as simply "one (social) thing" in "one (geographical) place", or indeed a.n.other. Nevertheless, in this context it is worth noting that in many cultural and even academic commentaries of surfing spaces, some degree of "insider" positionality is claimed by many authors (as just one example, see the excellent collection by Borne and Ponting, 2015). Often before any words of argument have been delivered, commentators and scholars refer to their positioning as somehow part of surfing spaces, this may be through shore-side habitation, first-hand (or first goofy-foot) experience, their ability on the water, their associations with famous individuals within surf culture, or their role in its development and mythology; and they do so in order to authorise their right to speak. It is as if the reader is expected to judge the significance of the author's status before they judge what they have to say.

The impression is therefore given that "insider status" is important when engaging with surfing spaces. Vernacular knowledge, gained through direct experience, is assumed to confer capital and credibility to any author from their audience. In many senses, this valorisation of experience is laudable in celebrating the embodied knowledge of surf-riders – these authors "know the feeling" (following the seminal Billabong tagline) because they have experienced it; there is little, perhaps even no, alternative for this first-hand engagement, as surfing space writer Allen confirms:

> the knowledge I speak of can only be obtained by increasing one's awareness of immersing ourselves in the environment. In no other way is this knowledge obtainable. To understand the ocean, we must be close to it, close enough to feel water, to smell the ocean breeze, to taste the salt that settles onto one's skin
>
> (Allen, 2007: 11),

and as surf scholars Olive et al emphasise:

> "... getting wet and getting involved reveal a depth of complexity and connection that cannot be gleaned from books, magazines and films alone, and allows for the dynamic physical and sensual experience of going surfing."
>
> (2015: 263)

or as Don Redondo, a "beach bum" from Malibu sums up:

> "You wanna know 'bout surfin', you surf... You don't surf, you don't know nothin' 'bout surfin'."
>
> (cited in Kampion, 2003: no page).

This elevation of the value of embodied practice suggests that there is a minimum sense of "literacy" required in order to be taken seriously when studying surfing spaces, a literacy based on first-hand involvement, and hard-won experience. As Olive et al go on to state:

> "Indeed, research into surfing and surfing culture has consistently spoken about the importance of participation as a surfer in order to gain access [and by extension, insight] to the culture"
>
> (2015: 263).

When discussing my own positionality, therefore, I consequently feel code-bound (see also Chapter Twelve) to conform to these established (b)orders of

etiquette and outline my own cultural credentials for studying surfing spaces – not in terms of being a "qualified academic" (although the transparency and accountability of academic positionality is in itself a well-established practice to maintain critical reflexivity and rigour, as discussed in Anderson, 2002), but as just another individual claiming a right to study surfing spaces. Indeed, in one sense I have fully complied with these (b)orders through offering an initial logbook entry in Chapter One (directly complementing Winton, 1993). As stated there, my earliest memories coalesce around family trips to the coast, and once I gained a privileged independence to explore the world, I extended my adventures from family swims, to boards, crafts, and (kayak)paddles with friends. As a result of both my personal and academic experiences, I would be tempted to call myself a "littoral being", someone who feels "just right" when I am part of any coastal waters, and as one who harbours a growing sense of absence and emptiness whenever I am away (see Chapter Four). These affective affinities are thus allied to and have in turn motivated over twenty years firsthand experience of participating in (and studying[1]) surfing spaces. And yet....
And yet, I still do not "feel" like I am an "insider" of surfing spaces; or better put, I do not wish to claim any conventional "insider" status. This may be obstinacy or awkwardness on my part – I am a more or less competent surf-rider, I am male, I am white, I am straight, I am blonde(ish), and I am half-heartedly athletic, so to some extent I do "fit" the dominant (b)orders of surf culture. Yet I do not in live in San Diego or Raglan, I have not competed at any level, I fit my surf-riding in and around commitments to family- and professional life, I would rather sit to ride (which may be a consequence of dodgy knees, or because its warmer to do so in the Atlantic winter; or if pushed, I might even argue because it's more fun (and it is! ?), or it may simply be because it has become "what I do"). Any combination of these identity positions mean I do not consider, and do not wish to be considered, a fully-fledged insider to surfing spaces. But I also realise that due to some degree of competency, a stubborn loyalty to and affinity with the practice, a personal curiosity and professional capability as a researcher, I cannot and do not wish to claim a fully "outsider" positionality either. So it is from this peculiar set of identifications that I claim a right to study surfing spaces; a position in-between the distinct duality of "insider" and "outsider".

A final turn towards edge-riding.

Reflecting many scholars who have sought to identify places before and beyond dualistic binaries (see for example Cloke and Johnston, 2005), this book studies surfing from a "thirdspace" position. Echoing neatly the in-betweenness in which many scholars define the physical materiality of surfing spaces (see Chapter Three), "thirdspace" identifies a liminal subject position between the conventional categories of insider and outsider for the researcher (see Soja, 1996, Routledge, 1996). In the context of surfing spaces, "thirdspace" is a therefore a site within, yet at the same time without, both surfing spaces and academic scholarship. As such, "thirdspace" not only acknowledges the diversity of political spaces that individuals inhabit, but also the fluidity and mutability of our subject positions. Using this approach, I am (now perhaps "we" are) able to "step into" my (our) repertoire of "selves" that inhabit academic spaces, as well my range of selves which occupy surfing spaces, and then step outside again. Claiming this vantage point means that in practice, "neither site holds [absolute] sway" (Routledge, 1996: 400) as academic and surfing voices do not silence, but rather continually subvert, support, and synergize with, one another[2].

Being sited in a space that is within and yet at the same time without both academic and surf-rider arenas is an important position to occupy. It at once means it is possible to recognize the importance of first-hand experience, but also be wary of any assumptions that could arise from this position. In the same way that Stranger's work on surf companies suggests that, from the perspective of surfers at least, only "insiders" could authentically and successfully produce apparel for surfers (2011), it may be tempting to suggest that it is only those directly inside the (b)orders of mainstream surfing spaces that can "know" about surfing, and thus legitimately write about surfing. By extension, it may also be assumed that it is only surfers who are interested in reading about surfing – or they may be the only audience that has significance for surf writing – as American surf journalist and editor Scott Hulet states, any surf writer "builds his reputation as a trusted voice to those who matter most; his fellow surfers" (in Taylor, 2005: sleeve notes). Such a perspective suggests that not only may a surf writer feel pressure to keep this key audience on side in order that their flow of (insider) credibility and capital is maintained, but also keeps the culture of surfing a relatively closed shop. A "thirdspace" position thus offers the opportunity to move beyond the constraints that insider surf writers may be influenced by, and as a result, lead to them (sub)consciously, "collude… with the reproduction of power-ridden values" which may "exclude… and silence" particular surf-riders and practices (after Pile, 1994: 255, 261, and see for example the gendered vocabulary adopted by Hulet, above). Thus the "thirdspace" position offers a working alternative to the complete insider or outsider binary, a position both within and without of surfing spaces, which generates an informed but critical perspective on their development and nature. Thus as Bakhtin notes, although

> there exists a strong, but one-sided and thus untrustworthy idea that in order to better understand a foreign culture, one must enter into it, forgetting one's own, and view the world through the eyes of this foreign culture… In order to understand, it is immensely important for the person who understands to [also] be located outside the object of his or her creative understanding – in time, in space, in culture. In the realm of culture, [this] outsidedness is a most powerful factor in understanding. We raise new questions for a foreign culture, ones that it did not raise for itself; we seek answers to our own questions of it; and the foreign culture responds to us by revealing to us its new aspects and semantic depths
> (cited in Marcus, 1998: 116).

36 *Turning towards Surfing Spaces*

This book is therefore generated from a "thirdspace" positionality, and it is from this location – drawing on the academic turns outlined in this chapter – that it offers its own "creative understanding" of surfing spaces. Aligning insights from cultural geography, tempered and weathered with first-hand experience of engaging with the cresting wave, this book continues by exploring the ways in which physical geographies assemble with human practices in order to create surfing spaces.

Notes

1 Indeed, my academic interest in studying surfing spaces has driven over two decades of research projects in this area. This book draws both explicitly and implicitly from the gathering of extensive secondary evidence throughout this period, from both contemporary and historical, scholarly and surf culture, materials. Over a dozen primary research projects have been undertaken during this period, involving a combination of qualitative and quantitative methods (including surveys, interviews, and ethnography), predominantly centred in the UK and Europe, but with many including Australasia and North America. Many of these projects involved a broader team, including Katherine Jones, Lyndsey Stoodley, and Matthew Walsh. These projects and archives, alongside my first-hand experience of wave-riding, inform my argument outlined in *Surfing Spaces*.
2 Perhaps pointedly, given the subject of this book and the hydro turn in human geography, working through thirdspace requires, "keeping intact one's shifting and multiple identity and integrity, [it] is like trying to swim in a new element… [it's] not comfortable, but [it becomes] home" (Anzaldua, 1987: preface).

Part I

Part 1

3 Locating Surfing Spaces

Introduction

We begin our exploration of surfing spaces by asking the key geographical question for any cultural activity: where is it taking place (see Anderson, 2021)? As stated in Chapter One, surfing spaces are conventionally defined as "oceanic" in their location, taking the place of the sea and making it into a surfing world. However, as this chapter will go on to argue, despite the shift from a geo- to hydro-logic approach within human geography (see Chapter Two), and this "watery" turn emphasising the importance of fluidity in both a material and philosophical sense (Anderson and Peters, 2014), there remains significant work to be done to create an appropriate "locational" language for surfing spaces. If, as Wittgenstein reminds us, "the world we live in is the words we use" (cited in Raban, 1999: 151), it is possible to move beyond the simplistic framing of water worlds as simply "blue" (see for example Buser et al, 2018), "wet" (Peters and Steinberg, 2015), or in the case of surfing spaces "oceanic" in nature, and better locate the complexity and significance of this realm. This chapter will seek to examine this complexity and significance; it will argue that surfing spaces can be better understood not simply in terms of their apparently fixed location, but more precisely in terms of the energies that transfer through this coming together of air, sea, and land. In order to therefore fully locate surfing spaces, the flows, origins, and futures of this energy require acknowledgement and, in doing so, this chapter argues that the littoral zone should no longer be simply considered as a "more-than-land" space, but rather a "more-than-sea" space, and due to the meteorological and ultimately solar processes at play, a "more-than-planetary" space too. Further developing the watery and relational turns we have seen in Chapter Two, this chapter moves beyond the limits of "flat", "wet" and "oceanic excess" discourses to argue that the phenomenon of surf is therefore a function of the littoral zone, a space which connects not only the here and there, but also the now and then. It is a spatial consequence of the meeting of land, sea, and air, of more-than-human energy flows, of material entities and fluid processes. We begin this reconsideration of the location of surfing space by casting off into the littoral from the land.

DOI: 10.4324/9781315725673-3

Locations from the land

As we have seen in Chapter Two, much human geography has been configured from the perspective of the land. Despite water worlds being integral to human life in all its basic, economic, cultural, and even spiritual forms, human geography has until recently been a resolutely terra-centric discipline. For example, if we turn to recent bibli(ographi)cal reference sources such as *Oxford Dictionary of Human Geography* and *Oxford Bibliographies,* which function to identify legitimate subjects for geographical inquiry, citations proliferate for the "city" and "urban" whilst there are none for "surf" or similar ocean-related terms[1]. In broader scholarly work in human geography, when surfing spaces are referred to a very general nomenclature is adopted. Scholars often site surf-riding activities as part of the land, as Shields (2004: 44) suggests, surf-riding activity is found on "the beach", whilst Preston-Whyte similarly states that "fishing, surfing,… boardsailing [and] scuba diving]" happen on "the beach" (2004: 351)[2]. Such generalization not only reflects but actively perpetuates the land chauvinism of geographical discourse as well as an ill-refined locational language with regard to surfing spaces[3].

Despite the terracentric pre-occupation of human geography, in some cases surfing spaces *are* defined in relation to water. In these cases, surfing spaces are often defined as "oceanic" in nature. Such examples are a positive step, they accurately suggest that surfing spaces are not "of the land" and encourage a different perspective on appropriate geographical study. However, as we have seen in Chapter One, it would be wrong to suggest that surf-riding commonly exists in all "oceanic" places all the time. Thus despite such definitions reclaiming the nomenclature of "water worlds" for surfing spaces, there remains considerable scope to critically unpack the variety of liquid bodies within this label (and as a starting point here, see Anderson and Stoodley, 2017). In sum, it is important to move beyond the generic locational language of the "ocean" to draw attention to the diversity of sites and zones in which surfed waves can occur. To this end, an important first step to is identify the original, and currently pre-eminent space of surf-riding within the hydrological cycle: the littoral zone.

Surfing the littoral

> Surf-riding occurs in the "transitional, highly porous border between the primeval terrestrial and aquatic"
>
> (Barilotti, 2002: 34).

Conventionally, the littoral zone is understood to be the place where the land meets the sea (or, depending on your orientation, the place where the sea meets the land). The littoral is geographically adjacent to both these elements, and due to this location, has come to be commonly regarded as the border demarcating these two key sites. However, as Barilotti suggests (above), this border is not fixed, stable, or impermeable in nature, rather it is moveable, dynamic, and "porous"; and it is due to these characteristics that the *littoral* has often been conflated with the *liminal*.

The idea of liminality emphasises the importance of the "in-between", drawing attention to the "threshold or margin at which activities and conditions are most uncertain" (after van Gennep, 1960, see also Hetherington, 1998; and Stronach and Maclure, 1997). For Turner, liminality exists where no one temporal or indeed spatial category holds sway, but focuses on the spatialized moment where two categories come together, where the dominance of either category:

> "is momentarily negated, suspended, or abrogated... [the liminal is a space] of pure potentiality when everything, as it were, trembles in the balance"
> (Turner, 1982:44).

Liminal spaces thus literally occupy

> "a threshold or borderland [sic], straddling, subverting, and disrupting boundary lines of separation and contact"
> (Jones, 2008:17).

With this in mind, the littoral-as-liminal is not simply a thin, fixed border which serves as a bulwark to purify both materially and categorically the elemental forces of "sea" and "land". Rather ideas of the littoral-as-liminal identify surfing spaces as a "third" site which "suspend" or even "negate" the characteristics, compositions, and assumptions surrounding these apparently opposed materialities. However, in general, the potential of the littoral-as-liminal has not been realized in contemporary examinations of the space. The dominant way in which the possibilities of the liminal have been employed can be seen to again reflect the terra-centric orientation of human geography. Littoral-as-liminal has been approached from the direction of the land, positioned as a margin to its centre, and perhaps not surprisingly, attention has been focused on the terrestrial element involved in the littoral-liminal: the beach. With this geographical and cultural space in mind, the littoral has been characterised as a place of temporary escape from the striations of the land, from the existing codes and controls of civilised society. The littoral-as-liminal has been understood as an area where people can live a (b)orderless life, beyond their terrestrially-oriented, "positions assigned and arranged by law, custom, convention and ceremonial" (Turner, 1969: 95). As a consequence, the beach is thus configured as a space of the "carnivalesque", where individuals can enjoy the "temporary suspension, both ideal and real" of the conventional (b)orders of cultural society (Bakhtin, 1984: 10). As Shields identifies, the littoral-as-liminal can be "spatialized [not simply] as margins [to the central land, but also] as 'free zones'" from that land's co-constituting rites and rituals (2004: 44).

As we will see in Chapter Ten, although the idea of the littoral as a "pleasure zone" for hedonism, sport, and sexual activity (see Hemingway, 2006) has been successfully employed to define the imaginary of surfing spaces, this reading of the littoral-as-liminal does not exhaust its potentiality. Indeed, as Preston-Whyte suggests, "the discussion on [the littoral-as-liminal] needs to be deepened" (2008: 356) with two key issues demanding attention:

> "Firstly, the human ... and the non-human actors that constitute the material conditions of the beach itself ... must be dealt with on equal terms. Secondly, dualisms, such as nature/culture [and we may add land/sea] ... need to be addressed"
>
> (Preston-Whyte, 2008: 356).

As Preston-Whyte implies, the dominant way in which the littoral-as-liminal has been approached bears the legacy of the terra-centric approach. As we have seen, the littoral is viewed from the land, with emphasis placed on the human geographies and cultures of that specific element, rather than the non-human physical agencies which contribute to "the material conditions of the beach itself" (and we may add, the broader littoral zone). There is a need, therefore, to take the "non-human actors" of the littoral zone just as seriously as the agency and effects of the human, and through doing so, critically "suspend" and even "negate" the dualisms of nature and culture in order to envision their existence in a new way. One way of achieving this is to approach the littoral-as-liminal from the sea.

Locating the littoral from the sea

If we take seriously the recent turns in human geography which have recast the land-scape (or perhaps better defined as element-scape) of the discipline (see Chapter Two), we can sensitize ourselves further to the nature of liminality by approaching the littoral from the sea. The natural sciences do just this; from these perspectives, the "beach" is not categorized as marginal to the land, but rather as belonging to the sea. As the *Collins Dictionary of Geology* suggests, the beach is,

> "the sloping shore of a body of water, especially those parts on which sand, gravel, pebbles or shells or other parts of marine organisms are deposited by waves or tides"
>
> (1990: 57).

From this perspective, the beach belongs to the water, it is created by and defined in relation to the sea. A simple turn to water thus immediately involves acknowledging the role of non-human agents in forming these spaces and invites the re-consideration of one's relation to ideas of "centre" and "margin", as well as the domination of fixity and fluidity. Echoing the consequences of the watery turn as outlined in Chapter Two, when you approach the littoral from the water, "you realise that the familiar world, the static world, of rocks and cliffs, is just one side of reality, and there is this other reality, where everything is [different]" (after Malone, 2011).

Thus in Preston-Whyte's call for us to explore the liminal more "deeply", we can begin by approaching the littoral from its apparent counterpoint, "the sea". However, our newly activated sensitivity to the power of language may also draw attention to the importance of "depth" in our understanding of the littoral, and by extension the limits of the land / sea binary. As we have seen, at first (horizontal) glance, the littoral zone may be accurately located "in-between" the spaces of land and sea. However, as second (vertical) glance, the littoral is not simply in-between these two apparent alternatives, rather it is the meeting place between *three* elements, namely the land, the sea, and the air. With these three elements acknowledged, we are at once beyond binaries and beyond two dimensions; from this perspective the littoral – and by extension the surfing spaces that directly concern us – not only require us to pinpoint a distance and time to understand their emergence, but they also require a "depth" and "verticality" to understand what apparently is at the "surface". When we take a view of the littoral from the sea, we therefore become closer to realizing that this is not simply a margin to the land, a "thin line" which demarcates categories which should be the focus of our interest; rather the littoral becomes a "centre" in and of itself, a space which due to its liminal nature offers alternative characteristics and compositions to land-locked ways of thinking. We can begin to explore the significance of this "new" space by focusing on the forces that generate the littoral-as-liminal.

The emergence of surfing spaces

> What many of us don't realize during our novice days of sea-sport activities is that the [littoral] is a very different playing field [to the land], and we don't have the home-court advantage. Skiing has a predictable hill and slope that stays somewhat constant. Water-skiing has some varying conditions, but it's still fairly much a sport of constant value when it comes to the Mother Nature factor. Wave riding sports, however, have many more variables to deal with
>
> (Cool, 2003: XI).

As Malone implies (above), the world as seen from the water is a world of movement; surfers glide, slide, or otherwise ride breaking waves. As a consequence, their "field" is not constant, it is most obviously liquid, but also influenced and constituted by other "formidable forces" (after Cool, 2003: XI). These forces come together to produce the motion we understand as waves.

> A wave might seem to be a simple thing, but in fact it's the most complicated form in nature. ...[As] one French scientist put it to me bluntly: "People have been studying waves for so many years, and we're still struggling to understand how they work"
>
> (Casey, 2011: 87 / 89).

Due perhaps to the multiplicity of "formidable forces" at play in surfing spaces, a full understanding of the generation of surfing spaces remains elusive. Yet to begin diagnosing how waves emerge, we must again acknowledge how views from the land can be partial, and views from the air misleading[4]; as Malone states:

> standing on the beach, we can only see a fraction of the ocean from where these waves originated. However, if we were looking from space, it would appear as though someone dropped a rock in the ocean, and ripples travelled out from its initial splash
>
> (2011).

Whilst views from the land limit our horizons and "views from the air" suggest that waves are moving like currents outwards from an "initial splash", water in the littoral zone does not behave like water in rivers. Water in the littoral zone is not turned, rolled, or transported by gravity downhill. Rather, in a similar way to a boat bobbing on the surface of the sea, water molecules move very little by the various forces they are subject to; as Pretor Pinney suggests:

> [When any] caravan of crests... roll in under a fishing trawler... [t]hey don't drag it towards the land, as they would... had they been currents of [river] water. ... The water that the boat floated on [and the boat itself] return to pretty much where it had started after the waves have passed through
>
> (2010: 13).

Considering the rise and fall of a boat as it is lifted up and down by waves serves to reveal the actual forms of movement at play in surfing spaces. Just as the boat rises and falls whilst waves pass under it, it nevertheless remains more or less where it started when the crests have passed. In the same way, molecules of seawater do not move outwards or forwards like a current in a river, rather they oscillate and rotate but nevertheless return somewhere close to their starting position. This process helps us to identify that water in the littoral zone can be mobilised without actual "forward" motion. Identifying this capacity means that when human surf, they are not simply riding molecules of moving water, they are riding water's capacity to "transfer" energy.

When humans surf they are riding the "formidable force" of kinetic energy. It is an illusion that the water which surfer's ride is moving outwards, the water molecules are in fact "simply" oscillating and transferring energy. Surfers therefore do not ride because water itself is moving forwards, rather they ride because water's molecular structure affords it the potential to be mobilized by kinetic energy. It is this combination of energy-and-the-capacity-of-water-to-be-mobilised that forms surfing spaces.

The energy that ultimately contributes to surface layers of water appearing to rise, crest, and break originates from many sources. Conventional perspectives may suggest that the key "force is usually, but not always, the wind" (Casey, 2011: 87)[5], however such lateral perspectives do little to recognize the generating force of the wind itself. As surf journalist Sharp confirms, it is important to acknowledge that "the sun is responsible for everything" (2014: 57). Although winds do generate waves (as we will see in detail in a moment), wind itself is caused by a change in temperature of the air. Energy from the sun is transferred to molecules of air, which in turn absorb this energy, become heated, expand, and rise. As this high-pressure air moves upwards, air of lower pressure rushes to fill the void beneath it. As this process continues, cooled mobilised air encounters the surface of the sea. Again, a transfer of energy occurs. The wind's kinetic energy is transferred through friction to the sea's surface and finds as it does so a new media in which to move. The surface of the sea now transports energy that ultimately originated from the sun, and waves emerge. Pretor Pinney describes this process as follows:

> The roughening of the surface [by the movement of cooler, low pressure air] increases friction between the water and the winds. ...tiny eddies develop [in the water surface which] result[s] in fluctuations in the pressure that the air exerts upon the water. The ripples... lift a crest here and sink a trough there and grow in size
>
> (2010: 19).

As these wind-blown ripples grow, they act like mini-sails (see Yogis, 2009) to further enhance the friction with the atmosphere, and cumulatively build wave size and speed. As Pretor Pinney goes on to describe:

> this confused and irregular ocean surface is called the 'wind sea'. Waves... begin to tumble over themselves, under the relentless, harrying force of the gale
>
> (2010: 20–21).

The force of wind is transferred from ocean molecule to molecule, in a kind of domino effect that reaches all the way across the ocean. If the body of water in which this domino effect occurs is large enough (in other words it has significant "fetch"), these tumbling radiations of energy "organise themselves into defined lines" (Fordham, 2008: 68) as the slower molecules become part of the faster molecules' train of motion. When this occurs, a "swell" has formed whose energy is now self-sustaining and no longer relies on an aerial force to propel it.

Thus, when we are beginning to grapple with the forces that generate the littoral-as-liminal we need to focus our attention not from the shore, or the skies, but out to sea. It is there we can identify the "formidable force" of the sun acting on the air, and then transferring this force to the water. It is from this perspective, over these time periods, and across these distances, that we can begin to identify the processes that produce surfing spaces in the littoral zone. As American surf writer Stephen Kotler sums up, it is with these realizations in mind that we can identify how the energy that roars towards us in the littoral zone is,

> quite literally a memory. It started out in some other part of the world, forming when a change in temperature produced a change in pressure. [As] air's natural tendency is to move from an area of high pressure to an area of low pressure [this] wind flickered across the ocean's surface, [and] produced small ripples which provided a greater surface area that can then catch more of that blowing wind. Eventually these ripples became larger and larger until they cohered into wavelets and eventually waves, attaining their greatest size when they came closest to matching the wind's speed. What makes this whole chain of events slightly stranger is that it is not the water itself traveling across the ocean as a wave, but merely the memory of the original wind's energy being constantly transferred as vibration from one neighboring water molecule to the next. When I hear the roar of that wave behind me at Nusa Dua, what I am actually hearing is the sound of the past arriving in the present with me directly in its path
>
> (Kotler, 2006: 23/24).

Going to sea thus affords us a new perspective on the here and there and the now and then, it also encourages us to consider how the surface of the sea may appear to be where surf-riding is at, yet it also reminds us that the emergence of this space is inextricably connected to processes occurring beyond and above the sea too. Yet when we consider the forces that generate the littoral-as-liminal, we need to do more than simply look out and up.

For more, see below

> "Q: On a big day you'll always hear some cowboy saying, 'Get out there, it's only water.' Is it only water?
> A: Water and reef"
>
> (Mark Mathews, Australian surfer, in Doherty, 2012).

As we continue to locate the relational forces which form the littoral zone, we note how the wind sea turns to swell as the energy passing through it accelerates. As Fordham identifies,

> "As the waves grow across the fetch, the waves' profile become greater and the circular particle movement of water moves deeper and deeper, below the surface of the ocean."
>
> (2008: 68).

It is vital to identify the expanding profile of a wave as it is through contact between this profile and the shore that surfing spaces emerge. At this moment of contact, the energy passing through the seawater encounters the stored-up energy of the ocean floor (or the surface of the submerged land, depending on your point of view). The formidable force of the land does not have the same capacity to transfer energy as water enjoys, and as a result some of the water's energy is absorbed in the land (and contributes to erosion over time), and some is reflected back to the water causing the wave's profile to change its shape. As the speed of swell at the base of the profile decelerates the energy in the water is pushed upwards. The circular oscillation of water molecules is concertina-ed upwards to such a degree that the energy wave becomes top heavy and "breaks". The manner of its collapse depends on the nature of forces in both elements at play: in general, a steep foreshore sculpts the water's energy into a tall pitching hollow wave, whilst a gently sloping gradient will crumple it into smaller tumbling waves.

Summary

Following the call of Preston Whyte to examine liminal spaces more "deeply", this chapter extends the locational lexicon of surfing spaces into a broader vocabulary of understanding. It introduces a new way of considering places which draws on existing insights, but also develops them. Firstly, it helps us to recalibrate the agency of non-human forces in forming the material conditions of the littoral zone. It suggests that water worlds do not simply offer habitat for non-humans, but when brought together with other forces, also enable them to realize new agencies. As Anderson and Peters (for example) state, the, "physical quality of the sea (in liquid form) makes it a mobile medium subject to the energies and forces of nature – the wind, jet streams, the extra-terrestrial gravitational pull of the moon" (2014: 10). As we have seen, the physical capacity of the sea has the potential to mobilise without moving, to transfer energy from the wind and ultimately the sun, through the kinetic motion of its molecules. This draws our attention not simply to the quality of the ocean, but the realization of that capacity when it is brought into relation with other entities and processes. Echoing insights from Chapter Two, it is these relations that must be the key unit of our analysis, as it is only through this coming together of apparently discrete components that particular capacities are realized. The "formidable forces" that Cool refers to above are only made manifest through these relations; following Ben Anderson, it is the force of these relations that is key, "their capacities to affect and effect, [and make] the difference" (2018: 1120). Thus to understand surfing spaces we must shift our attention to the "relational configuration" of which it is apart (after Ben Anderson, 2018: 1123); as assembled coming togethers, individual components remain exactly that until their potentials are realized through relational interaction. It is only then that the new space of littoral-as-liminal co-constitutively emerges.

An emphasis on relations therefore realizes the littoral-as-liminal not as a thin border which keeps two elements apart, or safeguards the rigidity of a modern mode of categorization, but one that realizes it potential to subvert and even negate a land-locked world view. In an important sense, therefore, the littoral-as-liminal is a new space which is "more than land", but similarly it is also a space that is "more than sea". When its generative forces are accounted for, it could also be argued that it is a space "more than worldly"[6].

As we critically unpack the liminal further, we can also identify that although surfing spaces seem initially preoccupied with the surface of things, a relational perspective suggests that they can only be understood by moving beyond a lateral framing of the world. The relational nature of littoral spaces means we have to move beyond a flat ontology (see Forsyth et al, 2013; Elden, 2013) and "challeng[e] horizontal approaches by opening up a vertical world of volume" (Steinberg and Peters, 2015: 247). This is most obvious when engaging with winds and extraterrestrial influences coming from on high, but as we have seen "what happens below the surface" (after Elden, 2013: 35) should not be understood as simply a metaphor (after Forsyth et al, 2013). Surf only realizes itself when the energy passing through water is transformed through contact with the land. Here unseen depths and submerged verticalities are brought into play, only then are the dormant qualities of energy-in-water activated as they combine into the new assemblage of the breaking wave. As Allen's eloquent summary outlines,

> the wave [of energy passing through the sea] is not a wave on its own but becomes so with the interaction of the reef. ...The ocean swells and the coral reef possess qualities. With the swells passing over the reef, these qualities become "interactive" in a sense that there is an exchange of attributes. Each quality is redefined. During this interactive process, each quality is no longer what it was. With the exchange of attributes, the original qualities become more than what they once were. It is not that the reef becomes greater because it is surfed over, but the interaction gives rise to new qualities that could unquestionably not exist without it. The phenomena lies in the relationship where the interaction is taking place. The wave takes on its particular form due to the reef that is passes over. And the reef is no longer simply living organisms sprouting new life and existence, but also becomes the "shaper of swells". These forces shape each other, for with each passing wave there are turbulent forces absorbed by the reef
>
> (Allen, 2007: 85–86).

We can see therefore that surfing spaces are more than land, more than sea, and more than flat; indeed, they also go beyond ideas of simply "wet" or excessively "ocean" (Peters and Steinberg, 2019). Rather they take seriously the consequences of considering material constitutions as part of wider processes (see Massey, 2006). What is crucial here is the (re)acknowledgement that each element is not acting in a vacuum, but rather they are "doing things" in contexts. Is it the interaction and interrelation with these contexts – and the other "doing things" in that realm – that generate and mobilise potentials into existence. From this perspective, the "new" space of littoral-as-liminal is not a world of fixed and immutable, but of transiency and ephemera. They are spaces of "precarious relational accomplishment[s]" (Philo, 2005: 824), only occurring through the coming together of a range of elements and forces-in-elements; surfing spaces are assembled not simply of materialities but of energies, they are not simply worlds of movement, but of worlds of transfer and flow.

These key insights suggest that we not only need to qualify the practice of surf-riding in the littoral, but also the nature of the littoral itself. Rather than thinking of our world as simply "land-locked" in nature (in both senses of this term), or indeed a water world (again, in both a philosophical and material sense), the initial outcome from focusing on surfing spaces suggests we need to attend to the world as emergent due to the generative force of contact, encounter, transfer, and change. Our attention is drawn to how ontological being and becoming is formed through the meeting of different flows, energies, potentials, and actualities which can be released, stored, and harnessed in multitudinous ways. We become sensitized to the ways in which our world is not a collection of objects, but rather a world of energy, its storage, release, transfer, and effect. This view focuses our attention on the ways energy mutates form and media, is transferred from now-gone assemblages into assemblages-in-the-making, and will be stored in about-to-be-formed assemblages in the future. This is a world of interactional re-composition(s), where all relations on different timescales create movement and form, objects and transfer, entities and processes, nouns and verbs. We may have to employ the inherited utility of modern ontological divisions to describe it, but we can also acknowledge that new categories may be necessary as such terms "lose their usefulness" (after Lakiri Dutt and Samanta, 2013: X) to identify the relations we are constituted through[7]. This littoral-as-liminal, this surfing space, is a world that brings together the there and then, but happens here and now; as Capp suggests, it is a space,

> where sea and land collide… when the energy silently coursing through deep water finally explodes upon the shore in a burst of white noise, [and] the eternal becomes the now. Every wave is a perfect expression of the present tense: it can't be grasped or prolonged, only ridden
>
> (2004: 115).

Focusing on the original and currently pre-eminent space of surf-riding within the hydrological cycle is thus an instructive exercise. It can further sensitize us to terra-centrism within current geographical thinking, with implications not only for how we consider centre and margin, but also the limits of binary framings altogether. This focus suggests that it should be the relational that is now the "foundation" of geography (contra. Peters and Steinberg, 2019). Surfing spaces are emergent, eventful processes that assemble temporally, dissipate and disassemble again, they are "a relational making simultaneously involving production, transmission, reception and interpretation through and within entities and materials" (echoing Revill's (2016: 245) work on sound). In turn, pertinent questions emerge as to how we, as relational individuals, collectives, and cultures are constituted in significant here and nows? How are we "imbricated" with other forces, and what effects do these forces have on the stability or mutation of other "things" with which we are assembled? Thus, this examination of the littoral location of surfing spaces sensitizes us to the capacity of the liminal to blend, graft, or transform our understanding; it offers valuable insights into the onto-epistemological nature of the world.

Notes

1 It is as if, according to these sources at least, these categories are irrelevant for human geographical study. By implication, it suggests that appropriate spatial scholarship should be focused on the agency and capacity of *terrestrialised* humans, as Clark sums up:
"critical thought has kept its focus firmly on the various achievements and potentialities of human agency ... by and large we have left the rest up to the physical sciences to sort out" (Clark, 2011: xiii; cited in Franklin, 2004: 5).
2 Such generalisations may not necessarily stand in for all these authors' identification of surfing spaces, as Shields clarifies, "Surfing is one activity that typifies certain beaches, *or more properly* the waves landing on them" (2004: 45, my emphasis); yet this short-hand conflation is significant and influences how we continue to identify and frame surfing spaces.
3 Although it is generally accurate to identify cultures of fishing, boardsailing or surfing as beach cultures, it is important to note that (unlike for example beach volleyball) the practices themselves are not terrestrial in nature. Although you can "sail" buggies on the beach (a practice known as land-sailing), you cannot board-sail (known in the UK and elsewhere as windsurfing) on the beach; you can only do that on the water. Scuba diving occurs *in* the sea (maybe in coastal waters, maybe the deep sea; but not on the beach); and whilst you may fish from the land, you fish *in* the sea.
4 As exemplified through a range of surf forecast resources, see for example https://www.surfline.com/
5 Other causes of waves may be more terrestrial in nature, either above or below the surface of the water, including for example earthquakes or landslides.
6 Indeed, as Chiaroni suggests, with these formidable agents in mind, we might consider the abandonment of any exclusive "earth bound' metaphors in relation to the littoral (see 2015), and as Peters advocates, the sea "should not be thought of as a discrete, independent, sealed elemental entity but instead one which is itself co-composed of a broader extraterrestrial assemblage" (2012: 1243).
7 Indeed, as Philo suggests, "in the ebb and flow of an energetic world, ...the purificationary conceit of the 'moderns' is revealed as exactly that, a conceit" (2005: 825–6).

4 Relating to Surfing Spaces

Introduction

As we know from the book so far, none of us is free from geography (after Said, 1993). Indeed, following the spatial turn across the social sciences, many scholars would suggest that we are all spatial beings now. In relation to littoral spaces specifically, the role of humans is crucial; if waves are to be surfed, then humans have to be involved – so the former can become surfed waves, and the latter can become surf-riders. The act of surf-riding therefore has the potential to change both the nature of a littoral location and the nature of the person encountering it – there is a generative relation not only between land, sea, and air which creates spaces which may be surfed, but also between the human and these geographies which create assemblages which are. The remaining chapters in Part I go on to examine the reciprocal relations between humans and littoral locations. They discuss how the potential for "tremendous complicity" between humans and place is realised through surfing practice, and how water worlds come to "thoroughly interpenetrate" many humans' sense of self (after Shields, 1991: 14). In aggregate these chapters make the case that it is not simply that humans should be considered as spatial beings, or perhaps even "geo-humans" (in order to recognise the role of geographical place in human's co-constitution), but also that surf-riders could be understood as "hydro-humans", constituted significantly by their relations to water worlds.

However, as Chapter Three has argued, it is possible to adopt a more sophisticated categorical vocabulary to understand the cultural geographies of surfing spaces. It is useful to move away from generic terminologies in order to recognise the different bodies of water that locate and influence human activities. More specifically, it is possible to identify that surfing spaces are not simply oceanic spaces in general, but littoral locations in particular. By drawing on (post)-phenomenological accounts which explicate the process of emerging relationality between people and place, this chapter demonstrates how different "scales" of water space can also come to be recognised as constitutive of the human condition. Initially, the chapter draws attention to the ways in which the term "water people" more specifically refers to saline space, in particular at the scale of sea or ocean, before identifying how littoral locations become

DOI: 10.4324/9781315725673-4

recognised more directly by surf-riders as integral to how they conceive of themselves as human beings. Finally, the chapter addresses how it is not just littoral locations in general that become influential in person-place co-ingredience, but that it is often specific littoral places (i.e. particular bars, reefs, or breaks) that become key to the generation and maintenance of human identity and well-being. Thus for many, the terms "water worlds' and "hydro-humans' become nuanced, multi-scalar nexuses of attachment and co-constitution for surf-riders, with different intensities of connection built up for different bodies of water in different places, through anticipation, experience, and practice. This understanding forms the basis for the extension and dispersal of these surf-riding relations in Chapter Five, when we introduce the role of riding technology into the surfing space assemblage.

Spatial beings

Many scholars argue for the importance of space to the human condition (most notably Massey, 2005; Soja, 1996, 2010). But what is the case for suggesting that geography plays a key role in making us who we are? If you viewed the world from a modern, scientific perspective, you could be forgiven for thinking that place and space were largely irrelevant to human behaviour. As Madanipour et al (2001: 7) identify, in many areas of decision-making the tradition has been to "treat space and place as unproblematic, as part of an obvious reality, often [simply] as a surface on which things happen". From this perspective, humans are outside and beyond geography, detached from the spatial context as if it exerts no effect on their actions. Nicholas Entrikin puts it this way:

> for most of the past century, geographers have approached their subject in a manner that could be characterised in terms similar to those used by Italo Calvino (1972: 77) to describe the fictional citizens of Baucis, who, as residents of a city built on stilts, lived in the sky and gazed at the earth through telescopes, never tiring "of examining it, leaf by leaf, stone by stone, ant by ant, contemplating with fascination their own absence"
> (2001: 694).

However, is it really possible to stand outside the world, when our feet, and our lives, are firmly planted within it? Indeed, isn't it this connection between ourselves and the places we live in that furnishes us with insight about ourselves and the world? The modern view of space is challenged by geographers who regard space and place as a medium for action rather than a container of it. From this perspective, geographies have an influence on the social actions occurring within them. As places are taken and made by different social groups, they become politicized and cultured by human beings, enabling some social life but constraining others. Places then, are not only a medium but also an outcome of action, producing and being produced through human practice.

The co-ingredience of people and place

> It is a mark of contemporary philosophical thought, especially phenomenology, to contest the dichotomies that hold the self apart from [...] place.... we can no longer distinguish neatly between physical and personal identity ... place is regarded as constitutive of one's sense of self
>
> (Casey 2001: 684).

As Casey notes, there is growing acknowledgement of the role that place plays in forming and influencing human identity. This acknowledgement stems in part from the exploration of perception through phenomenology (see also Chapter One) and been extended through geographical adventures "in the dialectic between people and places" (Davidson, 2003: 2) recorded by Tuan (1975), Relph (1976), Seamon and Mugerauer (1985), Sack (1997), Hillier (2001) and Preston (2003), to name several. In general, these works posit that the human condition is a profoundly spatial, or indeed *platial,* one, with identity both influencing and being influenced by its inhabited material places. From this perspective then, places produce and are produced by human practice, and it is through performative practice that human identity is re-articulated and re-fashioned (see also Bondi, 1997; Butler, 1990; Gibson-Graham, 1994; Rose, 1997). In Casey's terms, the human body encounters places (through a process he titles "outgoing') and simultaneously inscribes traces of location on the human self by laying down "incoming' strata of meaning (see Casey 2001: 688). As Maxey (1999: 202) points out, the shifting nature of self is "reproduced continuously through [this]... practice". It is through spatial practice that we "temporarily attach" ourselves to "subject positions" (Hall, 1996: 6), and it is through moments of spatial performance that human identities emerge. This set of practice-based co-constitutive relations is best summed up as follows:

> The relationship between self and place is not just once of reciprocal influence ... but also, more radically, of constitutive coingredience: each is essential to the being of the other. In effect, there is no place without self and no self without place
>
> (Casey, 2001: 684).

Thus, through practice humans and places become forever co- and re-constituted. Traces of our presence may be left on places through our footsteps in the sand, the construction of our buildings, or through our (un)intentional or ill-considered polluting activities (for example); whilst traces of place may be left on our bodies through the physical exertions needed to be in that environment (to paddle, to remain buoyant, to hike, or to breathe), but also in our minds as we reminisce, anticipate, or otherwise imagine past or future engagements with a particular place. As Casey sums up:

> Places come into us lastingly; once having been in a particular place for any considerable time – or even briefly, if our experience there has been intense – we are forever marked by that place, which lingers in us indefinitely and in a thousand ways, many too subtle for us to name. The inscription is not of edges or outlines, as if place were some kind of object; it is of the whole brute presence of the place. What lingers most powerfully is this presence and, more particularly, *how it felt to be in this presence*
> (Casey 2001: 688; see also Ryan, 2012: 56–57).

As Halbwachs (1992) argues, and Crang and Travlou (2001) concur, due to the co-ingredience of self and place, our co-constitution with places – and our memories of doing so – merge the physical, cultural, and the personal. Time alongside practice sediments meaning onto places, with personal memories meshing a location's cultural meanings on an individual and (potentially) societal scale. As Casey points out, "places [can] possess us – in perception, as in memory ... insinuating themselves into our lives" (Casey 2000: 199).

To acknowledge the co-constitutive role of people and place thus also reminds us of key arguments made by this book to date. Firstly, that the body is a key vehicle and register of engagement with the world, and goes "hand-in-hand" with recognising a role for the affective in "embodied" knowledges. This knowledge is gained though practice and physical engagement with particular places, with a range of tactility, sense, memory, and reflection creating a (geographical) "life" informed by "a spatio-sensorial embodied mind" (to use lisahunter's phrase, 2015: 41). Secondly, the notion of the "assemblage" is useful to frame these "lives" in places. As we have seen in Chapter Three, the notion of assemblage refers to a provisional "whole" "whose properties emerge from the interactions between parts" (Delanda 2006: 5). In the case of constitutive co-ingredience, the person–place relation can be considered an "assembled" whole or "assemblage"; a provisional entity that emerges from the "incoming" and "outgoing" interactions between humans and places. In this sense, as spatial and social beings, humans can be understood as having provisionally assembled, co-ingredient identities. Thirdly, and in sum, these insights remind us that the human condition is a profoundly geographical one. It is only through considering our relations to, in, and through place that we can orientate and come to "know" ourselves. By extension, these insights also suggest that "ourselves" are not a singular thing, that identity is not an isolated or essential entity. Identity, therefore, has now to be considered highly contextual – influenced by the cultures and geographies we are (literally) a part of. Echoing the insights from Allen (2007) at the close of Chapter Three, new engagements between components of any assemblage alter the qualities of all constituent parts, with new capacities being generated through this interaction. As a consequence, identity also needs to be considered as "relational rather than absolute" (Davidson, 2000: 642), with the fixed idea of a unified self giving way to an acknowledgement of dynamic, multiple and fractured identities (see Anderson, 2004)[1]. As a consequence of these insights, humans are profoundly spatial processes, not simply and abstractedly "human", but geographical-human assemblages, or "geo-humans": they emerge, diverge, dissipate, and mutate, dependent on the various spatial-social assemblages they are co-constituted by.

The missing element: "Just add water"

The vast majority of (geo-)humans lead land-locked lives (United Nations, 2017), and thus for many, water worlds may seem of marginal importance (as Duane states, "city dwellers know nothing about neap tides or the topography of local reefs for the same reason few Americans know a second language: not out of moral or personal weakness but because *it doesn't matter* [to them]" (1996: 5)). Nevertheless, just as many people do know a second, or maybe even a third language, the identities of many (geo-)humans are formed by their connections with territory beyond terra (see Peters, Stratford, and Steinberg, 2018). Water worlds play a significant role in shaping our selves, and vice versa (see Anderson and Peters, 2014; Humberstone and Brown, 2015). For many, physically "returning to water" (Taylor, 2005) in any form makes (geo-)humans feel as if they are (back) in their element.

Taken literally, the importance of water worlds to the human condition could be taken to mean *any* water, *any*where. Any proximity to or any engagement with water at any stage of the hydrological cycle – be it "wild", "domesticated", moving, fresh, salty, or still – could be deemed to have a positive effect on human's sense of self. It is suggested that these people may be termed "hydro-humans", those who feel more "at home" when proximate to water in any form. But, as we have seen, proximity is often not enough for these individuals; direct engagement is often required in order to feel "just right". The manner of this engagement may differ for each hydro-human (from swimming, surfing, sailing, diving etc.), but often leads to a development of a general experience and competency in this element. A popularised "catch-all" term for such individuals is "water people" (a term that is often unreflexively reduced to "watermen") and identifies how these individuals develop a high level of expertise through sustained and prolonged engagement in water worlds. As Mattos describes with respect to sea kayaking, the idealised caricature of the "waterman" [sic] is someone "with a total mastery of all oceanic endeavours" (2004: 8), and as Susan Casey sums up, a "waterman" [sic] could

> swim for hours in the most treacherous conditions, save people's lives at will, paddle for a hundred miles if necessary, and commune with all ocean creatures, including large sharks. He [sic] understood his environment. He could sense the wind's subtlest shifts and know how that would affect the water. He could navigate by the stars. Not only could he ride the waves, he knew how the waves worked. Most important, a waterman always demonstrated the proper respect for his element. He recognized that the ocean operated on a scale that made even the greatest human initiative seem puny
>
> (2011: 39).

Whilst we can acknowledge and seek to dismantle the patriarchal framing of the water*man*, we can also critically recognise in these descriptions the evaporation of all aspects of the hydrological cycle to leave water people engaging solely with saltwater. In these definitions, water people have expertise through engagement with the sea. However, in the spirit of attempting to create a more sophisticated vernacular of water worlds (see Chapter Three), it is important to note that there may be important differences in encounter and experience between saline and freshwater spaces, as well as the various "bodies" of water hydro-humans may be co-constituted through as a consequence. Indeed, perhaps in implicit reference to these distinctions, some scholars specifically adopt the term "salt" or "sea-water people" to acknowledge the importance of oceans to their geo-humanity (see Sharp, 2002; Robins et al, 1998); as Humberstone, for example, states,

> For sea-water people, as for me, the 'where' in 'wherewithal' is located in the fluctuating sea
>
> (2015: 36).

The term "salt" or "sea-water people" (see also McNiven, below, and Brown and Humberstone, 2015) acknowledges the embodied experience emerging from a relational conception of the hydro-human. As Casey's description above suggests, these individuals are assembled together with a variety of "components", including the non-human, the aerial, the energies and processes of wave formation, and with the role of the land; in other words, the term "sea-water people' refers to a fully emerging assemblage evoked by geo-humans coming together with these bodies of water. As Humberstone confirms:

> arguably for us 'sea people', we are subtly connected to and intertwined with the energies of the waves, the sea and the universe. Who we are, and what we become, is bound up within the dynamism of the sea
>
> (2015: 35).

As Zink articulates with reference to her own experience, there is something in engaging with salt-water bodies that aligns and feels apposite for her own:

> Moments when things feel right draw me back to the sea. The sea is a place where that feeling occurs in a way it does not anywhere else. Maybe what draws me back is that the assemblages that form at sea work to loosen the borders of identity and expand both perception and the capacity to enter into relationships with other bodies in new and interesting ways. I go to sea because it affords opportunities to be in relationships that "feel right"
>
> (2015: 81).

Although many identify their preferred assemblage with any saltwater sea space, as we have seen in this book so far, surf-riders identify with littoral locations as their particular hydro-geographies of choice. In the words of Kampion, this particular category of hydro-human feels most at home in the,

> dynamic give and take of land and sea, swell and tide. The people indigenous to this fluid landscape belong to the global tribe of surfers. They see the ocean differently than inlanders, differently too than the other fringe dwellers who seldom set foot in saltwater
>
> (2004: 1).

Surf-riders, therefore, feel their own sense of "just-right-ness" in littoral locations, where the sea meets the air meets the land, and where rideable waves break. This sense of co-ingredient belonging is well-described by Australian surf-writer and novelist Tim Winton,

> Down at the low-water mark, at the scalloped edges of the shore, the water is gigglingly cold. Clouds rise around our feet. The four of us hold hands and bend like a sail, raucous in the east wind, laughing with shock. …There is no one else around. I flinch at the sound of a school of whitebait cracking the surface a few metres away. It's alive out there. After the still, exhausted Aegean, where nothing moves but the plastic bags, it seems like a miracle. Call it jet lag, cabin fever, but I am almost in tears. There is nowhere else I'd rather be, nothing else I would prefer to be doing. I am at the beach looking west with the continent behind me as the sun tracks down to the sea. I have my bearings
>
> (Winton 1993: 4).

In such littoral spaces, surf-riders experience a new sense of what it is like to be human. Emerging through proximity and practice in the littoral, the human is now re-oriented with and through the meeting of land, sea, and air, and their sense of who they are is both dispersed and extended with these elements and other non-human life. As Steinberg puts it, this experience "may be specifically human, but it is an experience of something that both *exceeds* and *becomes* [ourselves]" (2015: xiii). Or as Australian scholar and surfer Clifton Evers puts it, littoral locations, "are part of our bodies and our bodies are part of them" (2010: 49).

The process of engagement with littoral locations thus generates a mutual interference of (non-)humans which transforms the surf-rider. For many, this transformation is most affective with regard to particular littoral locations, in other words, specific breaks, bays, or coasts. Australian surf-writer Fiona Capp suggests this affiliation with regard to Portsea, an area of coast she had surfed as young woman, and returned to after a long period of absence:

> "I had not realised that a body of water could have such distinctive contours, that it could feel so familiar after so many years. It was as if I had come home"
>
> (2004: 48).

Clifton Evers highlights how this co-ingredience is felt by the surfing body, and how this surf–shore attachment becomes part of the surfer identity:

> Surfers form a sensory relationship with the local weather patterns, [specific] sea-floors, jetties and rock walls. Surfers' bodies intermingle with the coastal morphology, and it can be hard to tell where the local's body begins and the local environment ends. …The environment and how it works becomes so ingrained that a local should be able to tell the different surf seasons by the way their body feels. We bond with the geographical turf
>
> (2007: 4).

Such bonding with specific littoral locations is not only articulated by "modern" surf-riders but connects contemporary, western practice with indigenous identifications in surfing spaces. As Bagshaw (1998: 156) suggests, Aboriginal Saltwater People (as described in McNiven 2004: 334) sense a "consubstantial identification" not only with specific features in particular littoral locations, "such as islands, sandbanks, reefs, rock outcrops, tides and currents, along with sea creatures" (McNiven, 2004: 332/333) but also the creator beings who made them. Engaging with these specific places through surf-riding enables contemporary indigenous peoples to become part of a broader spirit-scape of the Dreaming which connects homeplaces (to borrow Hooks' phrase, 1991) to historical fishing grounds and key coastal or island journeys[2]. As one Meriam Elder from the Torres Strait puts it:

> "I am part of the sea and the sea is part of me when I am on it"
> (George Kaddy, Meriam Elder, Torres Strait, 1999, cited in Sharp, 2002: 27, and McNiven, 2004: 329).

Animoto Ingersoll, an indigenous Hawai'ian surfer (2016: 1), articulates this co-ingredience through littoral engagement as follows; she states:

> "When I enter the ocean, my indigenous identity emerges. …I enter a process of reimagination as the power of the ocean continually reshapes me alongside the coastal shores of my home"
>
> (2016: 1).

For Ingersoll, this re-creation of her identity is particularly evident when she engages with the littoral through surf-riding; as she states:

> Hitting that first whitewall of water, I become a Kanaka Maoli (Native Hawai'ian) surfer… In *ma ke kai* (in the sea), my physical involvement with, and thus my physical capabilities in, the world evolve. I become more agile in the water than on land: I can soar, glide, dive, and spin. I'm faster in the ocean, and can better navigate coral heads than roads. …It

isn't until I enter ke kai for *he'e nalu* (surfing) that I am able to reconnect with my Kanaka heritage

(2016: 1)

For Ingersoll, it is through practical engagement that the human self disperses and extends, not only transforming its physical capacities and sense of ability, but also plugs it into a pre-existing network of cultural and spiritual co-presence, which is at once contemporary, historical, and always in the making[3]. Such identification in part reflects the "transpersonal ecology" of Warwick Fox (1990), where humans feel their "individuality" is more appropriately described by a co-constitution with not only familial ties (living and dead), but also the physical features of the land and sea in which they have lived. This interpenetration of the genealogical and the geographical is also evident in the Māori of New Zealand, where the Māoritanga identify with their *whakapapa* (broadly translated as kin, a concept which does not differentiate between human and non-human lives), and leads to a sense of *kaitiakitanga* (or guardianship) of people and place as essential to each other (see Wheaton, 2018, but see also Memmott and Trigger 1998, and Robins et al. 1998).

Conclusion

We have seen in this Chapter how the human condition is a spatial one. Many geo-humans articulate this co-ingredient spatiality in relation to water (and can thus perhaps be generically termed hydro-humans), and this identification is evident from the scale of the saline oceans (as "salt water people" or perhaps "ocean-humans"), through to littoral locations in general, and to specific littoral bays, breaks, and coastlines in particular (i.e. producing "littoral-humans"). It is important to note therefore that it is possible for an individual to emerge differently through affiliations at these different scales, and that these identity positions are not mutually exclusive or fixed[4]. It is only possible to attain a full sense of being a hydro-, oceanic-, or littoral-human in the moment of practical action (see Anderson and Stoodley, 2019). For individuals who aspire to be surf-riders, they have to engage with littoral locations (and thus become littoral-humans), and undertake the act of surf-riding. Regardless of any previous experience or attainment, each cresting wall of water offers an opportunity for the human to be (momentarily) realised as a surf-rider, and with each wave that breaks-unridden the potential of it becoming a surfed wave and the littoral-human a surf-rider is lost.

From this perspective, therefore, an individual is defined by the (spatial) act they are engaged in at any moment. In relation to surf-riding, every encounter with the littoral becomes a test or ritual for the human who wants to assemble a momentary surf-rider identity. If one does not position ones' body correctly, one (or one's many potential selves) does (do) not emerge as a surf-rider. If one wipes out (or crashes from a breaking wave), one does not emerge as a surf-riding body. From this perspective, even though we may commonly refer to individuals as "surfers", we are more precisely suturing an identity label to them that is historical in nature – they surf, or more accurately they have surfed in the past. In the current moment, unless they are actually engaged in the act of surfing, they are not surfers. This apparently pedantic exercise in linguistic accuracy is actually more important than it appears. It is significant as it draws attention to the momentary and elusive state that being a surf-rider involves; it reminds us of the spatial assemblage that is invoked through the apparently straightforward identity label of "surf-rider", the definitive coming together of appropriately assembled land, sea, and air conditions, as well as reminding us of the importance of embodied action: the imperative of correct positioning on the water surface, and the need for successful execution of combined experience and skill to ride the unpredictable and uncontrollable energy surging through the molecular structures of the sea. This surf-rider assemblage, the product of the unlikely convergence of sea, swell, solar energy, continental shelf, beach, wave, board, and rider, is elusive, but powerful, and when its state is attained its affect is felt strongly by the successful rider (as we will discuss in Chapters Six and Seven)[5].

As we have seen, this understanding of the relational surf-rider has little in common with the scientific and modern framing of a human, defined as detached and abstracted from their cultural geographies. As Lewis puts it, the modern or "metropolitan" body may be understood as "inorganic", "passive", "ocular", and "groundless" in their orientations (2000: 59), whilst more active, outdoor humans (what Lewis describes as "climbing bodies"), may be more "organic", "self-determined", "tactile", and "of the ground" in their orientation (ibid.). As Ford and Brown (2006: 122) recognise, surf-riders have more in common with Lewis' description of "climbing bodies", individuals whose engagements with the material world not only change the competencies of their human physicality but also generate new senses of connection and extension with the physical world. Yet although the "surf-riding body" has a "family resemblance" (to use Goffman's phrase, 1971) to the "climbing body" – specifically in terms of its relationality, tactility, and being defined by connection to its geography, it would perhaps be more accurate to modify Lewis' categorisations and suggest that all hydro-humans (including the surf-rider) are distinct entities to their terrestrial ("metropolitan" and "climbing") counterparts. Following the insights from this chapter, the "surf-rider" may be identified as: i) constituted by water space, ii) determined by the practical engagement of surf-riding, iii) informed by "a spatio-sensorial embodied mind" (after lisahunter, 2015: 41), and iv) defined in and through littoral locations (rather than simply "of the ground" (after Lewis, above). Indeed, any framing of the "surf-rider" should also include reference to the assembled extension of experiences that enable this state to be realised in the moment; this is a task perfected by Ingersoll in what she terms her "seascape epistemology", an approach to knowing the surf-rider and their co-constitutive relations with the littoral zone as:

> presumed on a knowledge of the sea, which tells one how to move through it, how to approach life and knowing through the movements of the world. It is an approach to knowing through a visual, spiritual,

intellectual and embodied literacy of the *aina* (land) and *kai* (sea): birds, the colors of the clouds, the flows of the currents, fish and seaweed, the timing of ocean swells, depths, tides, and celestial bodies all circulating and flowing with rhythms and pulsations, which is used both theoretically and applicably… for mobility, flexibility, and dignity within a Western-dominant reality

(2016: 5).

To me, such an epistemology resonates strongly with Barbara Humberstone's recognition (itself drawing Deleuze and Guattari (1981)) that humans are not,

"bounded entities" with more or less fixed identities, but [are] embodied mobile, permeable "propagating signals"… Who, what and where we become are constructions not only of social imperatives but also of the places, spaces and durations through which our bodies travel in time

(Humberstone, 2015: 36).

Such recognition comes closest to identifying the relational emergings that continually re-compose surf-riders; it is an onto-epistemology which, as Ingersoll states, "enables a reading and a knowledge of the self that resists the petrification of its own dynamic character. Identity is always plural and in continual recreation… [co-composing] ever-shifting and negotiating beings within a Western reality" (Ingersoll, 2016: 16). From this informed basis, Part I continues its relational examination of the emerging surf-rider by turning our attention to the various technologies which enable and in turn co-constitute the assemblage of surfing spaces; to this end Chapter Five explores the craft we use to ride the waves.

Notes

1 Drawing on the work of Jameson (1991), Featherstone (1995: 44) has noted that the concept of identity has become decentred, with the sense of a coherent, essentialised identity giving way to the notion of fragmented, malleable and often "multi-phrenic" identities. Maxey (1999: 199) argues that the postmodern condition results in there being "no fixed 'me' of which I am fully cognisant". Featherstone (1995: 45) concurs, stating that the way we understand identity has changed from "being something unified and consistent", to something "conceived as a bundle of conflicting 'quasi-selves', a random and contingent assemblage of experiences". Thus the postmodern conception self can be thought of as a notion that accepts that me, myself and I are not a unified, singular entity, rather a strategic and increasingly fractured one – or many, multiply constructed across intersecting, and often antagonistic, discourses and practices (after Hall, 1996: 4).
2 Such "consubstantial identification' is echoed in Bawaka Country, where the indigenous peoples suggest particular links to the land which cannot be described by simple "connectivity'; in their words, "It is beyond connectiveness. It is a matter of co-constitution. People… are made through Country, they are part of Country and Country is part of them. We all come into existence through relationships with each other and with the world itself" (Bawaka et al, 2015: 274).

3 Clifton Evers articulates a similar but perhaps more "western' re-'connection' when he describes his experience of surfing spaces:
   "We feel the world through our shivering muscles, flushed faces, raised eyebrows, throbbing temples, and the tingles down our spines. When my body feels something intensely, I really pay attention to what is around me. Feelings make certain experiences, places and stories stand out and embed themselves into our memory and habits. The immediate experience of a wave mingles with a thousand details from my past experiences. During a ride my body imports its past into its present glide, remembering past slides, rides, waves, reefs, wind, hoots and stories simultaneously as it negotiates the present direction, level, balance. and distance" (2010: 52).
4 Indeed, in Part II we will discuss how multiple identifications can exist in surf-riders which are not always complementary but can often be openly conflictual and contradictory.
5 Drawing attention to the performative moment of surfing also helps us acknowledge the absence of a surfer-body after its disassembly and exit from surfing spaces. Emphasising the performative therefore reminds us that many individuals who are co-constituted by and with water are not who they wish to be when they are away from seaspace; they experience a sense of exile from the element that has come to define them.

5 The Surf-Riding Cyborg

Introduction

As we have seen in Chapter Four, surf-riders are generated through practical engagement with water worlds and littoral locations in particular. This process of complicity and interpenetration between people and place realises new capacities within each of its constituent components and comes to define the co-ingredient nature of surf-riders and surfing spaces. Despite being characteristic of indigenous cultures and others who feel tribal integration with this "fluid landscape" (after Kampion, 2004: 1), this assemblage is not often acknowledged by modern knowledge systems, which prefer to retain an atomised, individualised conception of self. This chapter turns to another key component in co-producing surfing spaces which may be more familiar to the modern: the role of technology.

> Without [surf technology] there is no surfing participation [and no] subculture
>
> (Warren and Gibson, 2014: 10).

In this chapter, we will focus on the role of technology which, as Warren and Gibson suggest, enables surfing participation and creates surfing cultures. This technology is most commonly understood to be the surfboard but, as Chapter One has demonstrated, there are a range of technologies available to the ride the breaking wave. This chapter will draw upon the work of post-human(ist) scholars to argue that these technologies are a key "prosthesis" which "extends" and "disperses" the capacities of the "human body" (Hayles, 1999), enabling it to surf-ride in different waves and in different ways. Following Haraway, it will argue that these human-technology assemblages can be understood as "cyborgian" in nature (1991, 2016), they emerge as chimeras of muddled multiples including but not limited to what we once considered as (narrowly) "human".

The chapter will go on to discuss how these (geo-, hydro-, saline-, and littoral-)cyborgs (see Chapter Four), generated with and through different technologies, create the potential for different experiences and encounters with surfing spaces. They enable different waves to be surfed, and thus broaden and diversify the range of surfing spaces that can come into being.

DOI: 10.4324/9781315725673-5

We will note how each technology changes the "carrying capacity" of pre-existing spaces, meaning that different waves can be ridden at different times. Relatedly, each technology also enables different ways of surf-riding to be generated, creating different politics on waves, which are sometimes convivial, but often conflictual (and these new forms of culture will be discussed further in Part II). Each technology therefore produces a different "ontology" of surfing space, with different surf-rider identities and different co-constitutive cultural geographies emerging through practice. Each of these various ontologies enables new "relational sensibilities" to be produced through the surfing spaces assemblage, and registered in the surf-riding cyborg; and it is these emotions and affects will be examined in the remaining chapters of Part I.

Just "boards" and "bodies" (?)

In the popular imagination, surfing is often reduced to a singular and homogenous activity: in this caricature, surfing is (simply) board riding. However, in practice there are many approaches that can be used to ride the breaking wave. Surfers can be categorised in terms of those who like to lie to surf, those who prefer to sit, and those who must stand in order to catch their waves. For each bodily approach there comes an array of technologies that can mobilised (including the body, a board, or a boat); there are an array of styles that can be adopted (ranging from a "glide" to a "shred"), and an array of different waves that can be ridden (from steeply-curving barrels, to slow surges, to loose, crumbly waves (see Chapter Three)). With each technology, body position, and style, surfing spaces are changed. Surfing spaces are therefore subtly but importantly different for a surf-*kayaker* (for example) when compared to surf-*boarder*. The technology used is therefore crucial in generating a new surfing space assemblage. The chapter continues by introducing some of the technologies used to surf-ride, and the types of encounters produced through these assembled co-ingrediences.

Being carried by the sea: Body(board) surfing

For those surf-riders who prefer to lie down to catch their waves, bodysurfing and bodyboarding are the most popular surfing activities. Bodysurfing, sometimes known as wave sliding, is perhaps the closest physical encounter one can have with the surfed wave (for examples see Bodysurf, n.d., http://bodysurf.net/). Bodysurfing is the least mediated engagement with the surface swell, and involves floating then swimming before the cresting wave, and manoeuvring the human body (often using half-cut swim fins on the feet and/or hands) so the breaking wave can carry the human forward on its moving energy.

Bodysurfing thus involves swimming to locate oneself in front of a series of breaking waves, then timing your "swim-in" to catch a wave as it rises behind you. But, as Pretor-Pinney suggests, a surfer's body causes... drag as it passes through the water... bodysurfers will sometimes hold one hand out in front of them to act as a sort of hydrofoil. By planning on an outstretched palm like this, they can lift their torsos out of the water, reducing resistance and creating more speed

(2010: 297).

Due in part to the drag involved between body and sea-surface, body surfing is a tricky activity to master[1]. However, technological developments in the 1970s went some way to removing this "drag" and open-up lie-down surfing to new constituencies.

Bodyboards

The advent of the body (or boogie) board in the 1970s – a torso-length foam board upon which individuals can lie prone – minimised the friction between body and sea. Its relatively inexpensive cost, combined with its ability to be connected (most often) to the wrist via velcro tie and rubber leash, offered both buoyancy and safety to the novice participant, and made it a popular alternative to bodysurfing. Like bodysurfing, bodyboarding relies upon being well-positioned (initially with board and body facing to the shore as a wave approaches) and catching waves as they careen forwards. The bodyboarding encounter is well summed up by the American novelist and essayist Jack London, when reminiscing on his first experience of the sport when travelling in Hawai'i in 1907:

> I shall never forget the first big wave I caught out there in the deep water. I saw it coming, turned my back on it and paddled for dear life. Faster and faster my board went, till it seemed my arms would drop off. What was happening behind me I could not tell. One cannot look behind and paddle the windmill stroke. I heard the crest of the wave hissing and churning, and then my board was lifted and flung forward. I scarcely knew what happened the first half-minute. Though I kept my eyes open, I could not see anything, for I was buried in the rushing white of the crest
>
> (London, 2004: 23).

Although bodyboarding is sometimes seen as an entry level activity (see below), this underestimates the skills and capacities of the pursuit. Bodyboarding offers the potential to ride challenging waves and do so in style(s) that other technologies restrict (for examples, see ThreeSixty, n.d., http://www.threesixtymag.co.uk/; and Nemani (2015)). Indeed, Nemani identifies how, with a bodyboard, "a rider is able to enter a barrel more easily than a stand-up surfer" (2015: 94), and as Waitt and Clifton acknowledge, due to its unique capacities,

> bodyboarding has moved ... to surf "shallower" and "steeper" waves of both beach- and reef-breaks [which are beyond the scope of other technologies].... Very skilled bodyboarders can surf sheer wave forms that are

unsuitable for shortboarders, because they break too suddenly and sharply. Near vertical waves also enable highly-skilled bodyboarders to perform a number of "tricks" including the "drop-knee" (using one knee on the bodyboard), "airs" (going high into the air after "bouncing" off the lip of a wave) and "360s" (spinning around 360 degrees on the wave face)

(2013: 493).

Take a seat, go surf-riding

For those who prefer to sit down to surf there are a number of alternatives on offer. Surf-kayaking represents the modern equivalent of Polynesian canoes, kayaks, and catamarans, one the earliest recorded means to surf-ride (see Chapter Nine). Surf-kayaks offer a sit-in hull and require the use of paddles to generate the power and mobility to cross through the surf zone to beyond where the waves are breaking. Surf-kayakers can then choose which waves to paddle for, coinciding their position and speed to guide their route down the breaking wave. Surf-skis are flat(-ter) alternatives to surf-kayaks; here the would-be rider sits on the ski (rather than in it) often strapped with easy-release belts around the waist or thighs. Surf-skis are highly mobile when compared to surf-kayaks, but also highly unstable; sitting on rather than in the water they are liable to tip when stationary but will effortlessly glide when propelled by the sea's momentum (or your paddling). Sit-on-top kayaks, by contrast, sit deeper in the water, have more stability, and relatively less manoeuvrability. Often made from air-blown plastic, sit-on-tops offer greater control at stationary or slow speeds, but this stability trades-off the craft's ability to reach high speeds and prime mobility in fast water. Both surf-skis and sit-on-top kayaks are easy to "bail out" from when they do capsize – meaning that riders do not need to "roll", they can simply fall from their craft, and (as long as they do not lose it in the wave) can climb back on board when the crest has broken. All surf-kayakers are aware of the inevitability of rolling – of the need to test their capsize skill, find it fit for purpose, or find a way to survive an encounter with a breaking wave (or maybe more). The affective connection between the human, their boat, and the littoral is in part encapsulated by the following personal logbook entry:

> Then I enter my craft. Clothed in shorts, rash vest, wetsuit socks, snug in a neoprene wetsuit and corseted by a dry-suit top and buoyancy aid[2], there is a familiar comfort in settling inside my kayak cockpit. A mix of security and excitement, entangled with a tingle of apprehension and, if I'm honest, a hint of fear. The waves seen from a standing position seem negotiable, from a sitting position they appear daunting. You stretch the neoprene spray-skirt over the cockpit, making a waterproof fit, and sealing the deal that we are now one. A push from the hands as you balance the paddle across the hull, and you're away; released from the terrestrial shore, buoyant, grabbing for the paddle to propel a course through the soup.

The first splash of waves jolts you. The cold assault slaps you as the incoming swell breaks your bow and collides with your skin and eyes (see for example, Anderson, 2013c, http://youtu.be/MaJpPSwNNF0). You shake it off, smiling, sadistically glad for the intense shock, but certain that if you do not recover and prepare yourself quickly, the next one will demolish and drag you back to shore. In a kayak, riding up waves is often as fun as riding back with them. Where surfers may duck dive under advancing waves, kayakers ride them like a ramp, careering over the top, taking air for a moment as the wave passes beneath, then landing with a happy thwack on the leeside of the rise (see Anderson, 2013d, http://youtu.be/YLZTPK4XBTg). I remember Wilson capturing this feeling well: "I felt sparks of exhilaration beginning to kindle ... the entire front half would lift and crash downward as the kayak regained full contact with the water, and a quivering thump would run through my whole body. The surface of the sea was level with my hips, while the whole length of my legs, encased within the kayak, lay below water level. The result, part of the unique experience of the kayak, was to feel curiously part of the ceaseless motion of the sea's surface" (2008: 28–9).

I attempt to gather pace, timing my speed to coincide with the rising waves – if I'm too slow the wave will pass without me, too fast and I'll shoot ahead and miss the crest. It comes, it's upon me. I lean back, feeling for the drive. The boat rises, and with a pivot of my hips I can turn it easily along the lip, riding sideways and forwards towards the shore. Clean, fast. Hips pivot back, the boats wants to stay, but with a quick brace of the paddle it reluctantly straightens and instantly remembers the fun it can have being driven arrow-straight. A dozen runs later, twists left and right have been nailed, well nailed enough to feel as if incompetent is no longer an appropriate signature. And as I prepare my exit, just to remind me of my place, a wave collapses around me, dumping me onto the ocean floor, stealing my wind as down-payment and scouring my scalp into sand. Again I smile, glad to be part of it all.

Surf-riding through prosthesis

As we have seen thus far, technology is crucial in creating surfing spaces. Boards, boats, and bodies can all be crucial components of the surf-riding assemblage, which in turn engage with and become co-constitutive of the breaking wave. As in all the cases outlined above, when involved in the practical engagement of surf-riding, it is not often easy to clearly identify where the human body ends, and the riding-technology begins. This is most obviously the case when considering body-surfing, when boardshort, rash-vest, and fin become part and parcel of the body in its element; but is similarly ambiguous when the thin foam of body-boards become closely compliant with weight shift and body angle of the surf-rider, or -skis and -kayaks become belted or sealed into a co-constitutive body-tech unit. In these co-fabrications, the capacities of the human body are not what they once were. With the addition of technology, the "individual" has become extended and dispersed into an assemblage with surf-riding potential. In this way, the human body is akin to a cog, instrument, or tool which can be connected to other components to expand its capability. In our more terrestrial lives, this is commonly experienced through the use of spectacles or contact lenses to enable the improvement of our vision, the use of sticks to enhance mobility, or even the constant connection to cell-phones to improve our communication, resources, or sense of community. In these ways, technology becomes an extension of the human body and changes its capacity; as Johnson suggests, these technologies become "prosthetic augmentation[s] of the human body, enabling the body to *exceed* itself" (cited in Wilson, 2017: 137, emphasis added). When these technological prosthetics are enabled, the human is no longer a discrete, atomised individual, but rather a relational molecular being; we are not simply humans, but more-than-humans, we are all cyborgs now (after Haraway, 1991, 2016).

In referring to the human body as a cyborg, Donna Haraway emphasises how the rigid (b)orders demarcating the human form are not as fixed and impassable as we may be conditioned to believe. Haraway suggests that humans are fusions, assemblages of muddled multiples including not only what we once considered as narrowly "human", but also other technologies and things, bacteria, and viruses (etc.). As a consequence, the easy identification of where a human ends and the "other" begins becomes a messy, monstrous exercise, and new capacities emerge through these cyborgian chimeras. The surf-rider, extending and dispersing the human capacities through technologies such as boards and boats, is a quintessential example of the cyborg. Technology regenerates the human body into an amphibious state through the co-ingredience of drag-reducing, gravity-resisting, prostheses. These now more-than-humans have the potential to ride breaking waves in new and different ways, and extend the geographies and experiences it is possible for these bodies to encounter. We can see further examples of this cyborgian "ontology" by moving on to discuss the range of surf-boards used to ride the energies of the littoral zone.

Stand up to be counted

For those who wish to stand up to surf, the dominant technology is the popularly known surf-board. Boards come in a range of sizes with different aims of mobility and speed; they may be broadly categorised as Stand Up Paddle boards (or SUPs), long-boards, and short-boards. Stand Up Paddle boards are long, wide, and deep "boards", often made from wood, polyurethane, or even inflatable plastic. In contrast to other boards, users stand not only to ride waves but also to move from the shore and navigate the sea itself; this is accomplished through the use of an extended paddle. These paddles help to generate momentum as well as steer the moving SUP. The use of Stand Up Paddle boards was documented in pre-colonial Hawai'i, and this technology has experienced in resurgence in popularity in the twenty-first century.

Due to their paddle and the rider's up-right position in the waves, stand up paddle boarders not only have great manoeuvrability to wave's peaks, but also can see where the optimal position may be to catch them (as they are not prone to paddle with their arms, or sitting to float and wait). As Laird Hamilton (a widely respected "water-person" and strong advocate of SUP innovation) suggests, stand up paddle boarders are not limited by the "worm's eye view" of other surf-riders (Casey, 2011: 113); due to their ability to be upright, "standup riders [can] see clear to the horizon[,] they identif[y] the best waves early, and then use… their paddles to accelerate past any other takers" (ibid.[3]). This technologically-generated ability means that SUP-users have an advantage to catch waves over other aspiring surf-riders. In terms of their subsequent surf-riding style, due to their size SUPs exhibit some similarities to long-board capabilities; as Fordham suggests, SUPs are "ideal for easy wave-catching and smooth-flowing, style-conscious manoeuvres" (2008: 75).

Long-(surf)boards may have similar lengths to SUPs, but are generally thinner and narrower than the very broad and thick Stand Up Paddle boards. Long-boards are ridden predominantly in a standing position, and are popularly linked to the development of modern surfing through the marketing of Hawai'ian cultures by those from California in the 1920s and 30s (see Part II). Long-boards, due to the length and weight, encourage a slow, sedate surfing style, allowing the rider to move along to the front of the board to "hang ten" (i.e. place all their toes over the nose of their board as the wave carries them). This board and style is commonly perceived to have an aesthetic elegance and due to its cultural connections to Hawai'i, alongside the craft required in manufacturing long-boards to suit the individual preferences of rider and littoral location, is a technology often associated with so-called "soul surfers" (see for example Fordham, 2008: 162; Warshaw, 2005: 552). Perhaps somewhat paradoxically, the largest and most durable long-boards are known as "guns", and rather than being made for their stylish use on slow, shallow waves, are made for their durability and strength at withstanding very large seas[4]. Variations on the long-board are used on some of the largest, offshore waves, where they and their riders are towed-in to rising surf up to 50 feet (Warshaw, 2005: 643). In smaller waves, long-boards can be crafted to suit individual riders and conditions. As British board-shaper Steve Croft notes, long-boards can be designed to,

> flow with the wave and give a feeling of glide. They maximise the energy in the wave by using modern rails and rockers, but with alternative templates and fin combinations to give effortless speed with the drive to use it. The longer curves smooth out twitchiness encouraging a more graceful, flowing style and promote more powerful, controlled turns
>
> (2008: 100).

The widely-regarded American surf-writer Drew Kampion introduces us to the experiential world of this tactile, embodied long-board culture. He begins by describing the importance of the ritual of waxing your board (in order to aid friction and grip in the transition to standing), before entry to the water, and riding the wave:

> Rub the wax in figure eights onto the deck of the board, stick it in your pocket. Run down the beach, feet sinking an inch or two, the sand moist and grainy and warm where it is wet with the ocean. You come to the rip, wait for a wave to rush up to the crown of the sand, then run with it as it flows back to the sea. Throw the board down, out ahead of you, and slide onto your belly. Paddle as fast as you can till you are outside of the shorebreak, heaving and crashing and splashing out with the rip. A quarter mile in three minutes and you are working your way out of the rip and over into the line-up. Some friends are there, wet already, and they greet you with big smiles. The waves are perfect, one says. Another yawps agreement... You are flushed with the run and the paddle-out, and exhilarated by the water
>
> (Kampion, 2004: 22).

Welsh surfer and writer Tom Anderson continues to describe this particular surf-riding experience as he finds a wave upon which to emerge with his board as a cyborgian surf-rider:

> after slipping down a solid wave myself and scraping to hold on, I found myself careering towards an oncoming section. The power under my feet was a rare sensation – that bliss of feeling the perfect line on a wave that is stacking up exactly as you want it to. Blessed for a moment with the exact sense of what to do, I jumped down from a floater to land metres from the shore, kicking out just in time to avoid running aground. Behind me a lull between waves had temporarily calmed the sea. I paddled back out, energised by the feeling I'd just had – the wind in my hair as I raced across the wave, the responding board under my feet, water drawing up the wave face, transferring its power through my body
>
> (2010: 40).

The aesthetics, style, and wave types associated with long-board culture were overtaken by the advent of the shortboard.

Techno-revolution

> In a single year [1968], the sport was almost completely transformed. Surfboards went from 9'6" to 8'6" to 7' and below, and anyone on a longboard was surfing a dinosaur
>
> (Kampion and Brown, 1997: 111).

Technological developments in the late 1960s enabled surf-boards to be made more quickly, cheaply, and of lighter materials[5]. These innovations in weight, design, and manufacture also enabled board-makers to innovate with (more mass-) production techniques and make lighter, shorter boards which were more affordable to the public. This combination of advances was key in promoting surf-riding in the 1950s and 1960s, initially in California (see Part II). Short-boards also enabled new styles of riding to be pioneered; lighter and shorter boards transferred greater speed and agility to surf-riders, releasing them to "turn more radically and to take aerial manoeuvres out over the lip of the breaking wave" (Fordham, 2008: 154).

> *Off the Lip* (1971) made the cover of *Surfer* that October and heralded a new frontier in the shortboard revolution, and the surfer's first step into space
>
> (Severson, 2014: no page)

In this quotation the surf-entrepreneur and -artist John Severson not only demonstrates his ability to imagineer surf-riding through connecting it with key American idea(l)s[6], he also identifies how the short-board changed surf-riding styles. With the advent of the short-board, riding spaces became less about the slide and glide, and more about flips, aerials, and shreds. The performance of boards facilitated the merger and mutation of techniques from other youth cultures such as skateboarding with surf-riding, with new styles such as "hacking turns and smashing lips" increasingly prevalent (after Croft, 2008: 100). These more aggressive styles lent themselves to different types of waves, as Fernando Aguerre (the President of the International Surfing Association) states,

> "Shortboards require a certain quality of wave to have fun. You can't really have fun on a one foot wave on a shortboard."
>
> (2015: 35).

Aguerre goes on to elaborate on the new frontiers in wave types that opened-up due to short-board technology; due to the capabilities of these boards:

> waves are no longer just a place to go gliding and go from top to bottom, they are now also used as a ramp. To perform the greatest air reverse you don't really need a great point break, a hollow wave that closes out after two manoeuvres is a great ramp for flying
>
> (ibid.: 36).

The short-board therefore enabled new styles of surf-riding, and new waves to be ridden. New short-board styles, with their possibilities for tricks and manoeuvres also opened up the potential for surf-riders to be judged on their execution of novel exercises and turns, and as a consequence, new impetus was given to other surf-riding reinventions as a regulated, institutionalised, and judged form of competition (see, for example Pill, 2019).

Conclusion

Development in board-riding technology thus significantly changed the nature of surf-riding and surfing spaces. The range of SUPs, long-, and short-boards, alongside body-boards and a range of surf-kayaks, demonstrate how a range of technological prostheses can all be used to generate the surf-rider. Through these prostheses the surf-rider emerges as a more-than-human-cyborg, an assemblage re-constituted with the capacity to ride waves. To paraphrase Haraway, all surf-riders thus become "chimeras, theorized and fabricated hybrids of machine and organism; in short, we are cyborgs" (2016: 7). Adopting this vocabulary is an important exercise. Recognising the technology that re-generates the potential of the human focuses attention on the different types of craft that co-compose surf-riding, the various waves that can be assembled as part of surfing spaces, and the range of styles of surf-riding that can be adopted. In other words, it sensitises us to the ways in which, as this cyborg goes geographic, new surfing space "ontologies" are developed. As Warren and Gibson implied at the start of this chapter, these technologically-mediated ontologies are vital to the development of surfing spaces, as each come to value different styles, waves, peoples, and places in different ways. To paraphrase Haraway, it is these (geo-, hydro-, saline-, and littoral-)cyborgian assemblages which are "our ontology; [they] gives us our politics" (2016: 7). As we will see in Part II, as each assemblage composes its own valid iteration of surfing space and diversifies the range of co-productive geographies, cultures, and experiences that are possible, these ontologies become the basis for conviviality and conflict between the differing emergent subcultures of surfing spaces.

As we have seen in this book so far, surfing spaces are not fixed or stable "entities", but rather are assemblages, composed by the variable and generating capacities of "relational settings" (after Somers, in Brubaker, 1996: 18). Although each of these relational assemblages is unique, what they have in common is the capacity to transfer the energy of these spaces to the surf-riding cyborg who is co-constituted by them. The nature and power of these new sensations and affects – of these relational sensibilities – will be examined in the remaining chapters of Part I.

Notes

1 I experienced the difficulties involved in body surfing during a trip to the Sea of Cortez, as I recounted in a logbook entry:
"We had spent the morning sea kayaking from an island in the centre of the Sea of Cortez, Southern California. We had paddled across flat water to a remote bay on the peninsula of Baja, a few hours south of Loreto. We had covered the distance in three to four hours, racing ahead of the winds, which created a surface chop on the sea. By the time this chop had reached where we were camping that evening, we had set up our tents and the wind's energy had coalesced the sea into a series of waves. This new shore-break was too inviting to ignore. We had no boards with us, and the sea kayaks were too long, unwieldy, and packed with provisions to ride such surging swell. I decided to bodysurf the waves. The waves were regular but their height and speed made timing my swim-in difficult. I had no fins and so had to kick

and paddle like fury in order to generate any sort of speed, especially compared to that produced by the swell. I also had to compete against my lack of natural buoyancy and the absence of a wetsuit, which would trap some air and give me a 'lift'. Nevertheless, with a sandy shore and enough collapsing froth to tamper the energy of the broken waves, the risk of serious impact was low. I was therefore free to clumsily 'porpoise' my way through the waves, being lifted and carried by the crests, my feet rising first, then being shot forward as if down a supercharged children's slide".

2   The need for all this kit, especially in cold-water surfing areas, also renders surf-kayaking less accessible and thus less popular.
3   As self-titled bodysurfer/waterman Noa Markou identifies, some refer to SUP use as "stand up snorkelling" due to the ability of these surf-riders to see marine life whilst they are active in the littoral zone (in Drift, 2007: 33).
4   As Jeff Bushman, board-shaper from Hawai'i states, there are large waves that: "are moving so fast there is no way you can generate enough speed paddling on a small board to get on to them. The Waimea wave will just pass right under you or it will throw you [for these waves, you need longboards]" (in Pretor-Pinney, 2010: 307). As Pretor-Pinney explains, the boards you need for these waves are "known as 'guns', [and] are at least 3m long. Since they spread your weight over a larger area, they are more buoyant and so have less drag" (ibid.).
5   These technological developments also meant surfboard were made of more environmentally damaging materials, both in the stages sourcing, manufacturing, and disposal. These have only been raised significantly as issues within surfing culture in the twenty-first century, but still remain marginal to dominant debates (see Part II, and Borne and Ponting, 2015, for example).
6   Namely, the conquest of the frontier and alignment with the contemporary zeitgeist of cultural and political revolution (including the success of the space race). This transformative ability of Severson and other key surf-entrepreneurs will be discussed in more detail in Part II.

6 The Relational Sensibilities of Surfing Spaces

Introduction

So far in Part I we have explored the "where" of surfing, and how these littoral locations influence surf-riders' sense of geographical self. We have argued as a consequence, that surf-riders ought to be understood as spatial beings, specifically as littoral-humans, co-ingrediently constituted by their affection and connection to surfing spaces. We have explored "how" humans ride waves, utilising a range of different technologies to enable them to glide, slide, or otherwise ride the energy passing through a littoral location. We have seen how the employment of these technologies adds new components to the geo-human assemblage, with each technology generating new capacities to ride different waves and execute different manoeuvres. In this way, each "littoral-human" becomes a "surf-riding cyborg", extended and enabled by technological prostheses to emerge through new encounters and experiences.

As we have seen in Chapter Five, each surf-riding cyborg assemblage is unique: different "littoral locations", "humans", and "technologies" – all already assemblages of various components – come together to create different ontologies of surfing spaces. In the chapters that remain in Part I, we will explore how the (now broadly-defined) "human" component in these surfing spaces chooses to articulate their experience of emerging through these unique assemblages. Firstly, in this chapter, we will argue that despite the range of experiences emerging through the different ontologies of surfing spaces, it is possible to identify some broad commonalities in these articulations. These similarities do not necessarily or inevitably emerge in a predictable fashion (there is no obvious cause and effect/affect). In other words, if one rides "this" wave with this "technology", it is not evitable you will experience "this" affect; this is due to the ontological reality of surfing spaces: there is no "this" wave (it breaks but once, and then it is gone), and its nature in the moment is generated in part due to how it is engaged with and performed by the -rider (themselves not a unitary or essential thing), and the manner in which their skill is executed in that encounter. In a similar fashion, the generating potential of the technology also changes through its use and elemental engagement. Despite or perhaps because of this ontological emerging, what is possible to identify is that

DOI: 10.4324/9781315725673-6

through each assemblage feelings and thoughts are generated within the surf-rider. Developing the vocabulary adopted in this book, and drawing on work published elsewhere (Anderson, 2009), this chapter argues that these experiences can usefully be understood as "relational sensibilities". Relational sensibilities are the emotions felt within the surf-rider but produced through the co-constitution of that human within a broader relational setting, in this case, the surfing space. The chapter will argue that these relational sensibilities bring forth a new awareness or attention of the world for the surf-rider, notable for its difference to that which they experience in more terrestrial locations.

The chapter will use Seamon's concept of "attentiveness" to explore this relational sensibility, before moving on to discuss key aspects of this affective awareness (1979). Before Chapter Seven examines the dominant lexicon with which surf-riders describe these experiences and affects – the language of "stoke" – we will first look at three key aspects of this relational sensibility as broadly noted by surf-riders: namely, "flow", "convergence", and "littoral religion". Flow refers to the mental state which is generated in and acknowledged by surf-riders through engagement in surfing spaces. Flow is achieved when surf-riders are fully "in the moment" in terms of their concentration, so much so that it is possible for them to execute (high) performance moves. "Convergence" refers to a relational sensibility in which surf-riders no longer sense the limits of their physical embodiment, but rather feel as if they have become part of the broader surfing space assemblage. The third relational sensibility, "littoral religion", connects to and extends the state of convergence and refers to how some surf-riders sense transcendent, spiritual experiences when on the waves. Complementing broader identifications of "aquatic nature religions" (Taylor, 2007, 2008), littoral religions refer to those relational sensibilities which move "convergence" experiences into the realm of "comm-union".

Relational settings and relational sensibilities

As we have seen in this book so far, surfing spaces are not fixed or stable "entities", but rather are assemblages, composed by the variable and generating capacities of "relational settings" (after Somers, in Brubaker, 1996: 18). Surfing spaces thus emerge through encounters which are transitory, ephemeral, and fleeting. Despite this ephemeral nature, surfing spaces have the capacity to be transformative, they change the capabilities and properties of the components that are temporarily part of their composition (see Chapter Three). This transformative capacity can render surfing spaces more durable in nature, or put another way, it makes it possible to witness or experience their outcomes long after the wave has passed. Indeed, surfing spaces can be more durable through "human" involvement in and through them. Our co-constitution with surfing spaces has the capacity to transform us, with ongoing effects on our motivations, mobilisations, and our ways of looking and being in the world; it is the case that relational settings generate relational sensibilities.

As stated above, relational sensibilities are the emotion felt within a human being but produced through the co-constitution of that human within a broader relational setting, in this case surfing spaces. Relational sensibilities therefore do not derive from a practice in isolation, nor the chunk-to-chunk practice of a person and a place, but rather they emerge through the practice-in-this-coming-together of unstable components (namely: non-humans, biospherical processes, culture, environmental context, etc). A relational sensibility is, therefore, part and product of the assemblage of any (littoral) location. By giving weight to these relational sensibilities, surfing spaces come to be defined not only by the coming together of non-humans, humans, and place, but also by these emotional responses; surfing spaces comes to be co-defined by "how it makes us feel". To be precise, however, a relational sensibility does not refer to "how I feel in the face of…" (after Game, 1997), but rather it refers to "this is how I feel being co-constituted by… ". It through being merged within a surfing spaces encounter that a relational sensibility arises.

Encounter and attentiveness

As we have seen, surfing spaces are moments of encounter. According to the humanist geographer David Seamon, an encounter is "any situation of attentive contact between the person and the world at hand" (1979: 99). In this view, the human is conceptualised as a receptive component of the assemblage of which they are a part, and as Seamon suggests, this capacity for attentive reception can have two broad characteristics: a "tendency towards separateness", and a "tendency towards mergence" (Seamon, 1979). It may be posited that during many encounters in our lives we experience a "tendency towards separateness". On these occasions, our cultural geographies are present and salient to us – they demand our attention to some extent, but we may feel ambivalent or otherwise unattracted to them. Indeed, the assemblage of which we are a part may be distracting to us, it may interfere with or disrupt our focus on specific component relations that we wish to attend to; we might experience this "tendency towards separateness" as our cell phone pings during a face-to-face conversation, or as the kids argue when we need to focus on work[1]. However, the "tendency towards separateness" can also be experienced in a more positive way. Seamon suggests that when a human concentrates solely on a singular practice, a separateness can be sensed from the broader assemblage of which they are a part; this can occur to such an extent that the assemblage becomes less salient, visible, and obvious to the receptive human component. In these circumstances, the assemblage has ceased to be a potential distraction and seems to disappear into the background. As it does so, the situation emerges where the human can fully become "lost" in their specific task, and as a consequence, the sensation is experienced that time has "stood still" (after Seamon, 1979).

As a broad counter-orientation to the "tendency toward separateness", Seamon identifies the "tendency towards mergence". When experiencing the

The Relational Sensibilities of Surfing Spaces 77

"tendency towards mergence", the cultural geographies of a particular assemblage are not ignored by the human receptor, rather they are fully sensed and acknowledged. However, in these cases the geographies are not experienced as distracting, distinct, and unattractive (as in the case of "separateness" outlined above), rather they are sensed as attractive and appealing, connecting and extending aspects of the human in desirable ways. Indeed, in these states, the human again "loses themselves", but rather than their attentiveness being directed towards the self and the performative task at hand, it is directed outwards towards the myriad of component relations they now feel a part of. In this "tendency towards mergence", human receptors sense a re-generation as part of something larger and greater than their isolated selves, and this co-emergence is welcome and positive to them. As Seamon suggests, these "tendencies" are precisely that – possibilities and potentials that may fluctuate and be impossible to predict in their regularity or intensity. However, this chapter suggests that these tendencies do become useful parameters with which to discuss the relational sensibilities that surf-riders articulate through their "encounters" with surfing spaces. Here we will specifically note how a "tendency towards separateness" usefully couches the experience of "flow" that many surf-riders identify and seek when engaging in surfing spaces, whilst the "tendency towards mergence" contextualises firstly an experience of bodily convergence with the breaking wave, and secondly, with a spiritual communion many surf-riders sense as a consequence of their littoral encounters[2].

"Flow" experience and the surfed wave.

Many surf-riders articulate a tendency towards separateness when engaging with surfing spaces. This tendency is discussed through the psychological term "flow" (as noted and discussed by Shields, 2004: 50; Ford and Brown, 2006; and Stranger, 1999). Flow is a term coined by the participants in a psychological study conducted by Csíkszentmihályi (1990), and originally refers not to a sensation based on engagement with water (as one may anticipate), but rather to any state of mind in which an individual feels "in the zone", so much so that they can successfully direct their full concentration to a specific task. The flow experience is defined by Csíkszentmihályi (1990: XI) as an "optimal state of experience in which an individual feels cognitively efficient, deeply involved, and highly motivated with a high level of enjoyment" (also cited in Asakawa, 2009: 123). As such, "flow" has been used to refer to mental states emerging through a range of activities including playing tennis, skateboarding, writing, mountain climbing, meditation, playing chess, or creating art.

As Rogatko (2009) outlines, the attainment of a flow state is dependent on the setting of rational goals to direct whatever activity is being performed. For flow to be achieved it is vital that specific objectives are clearly demarcated, targets which are within one's capability, but will nevertheless demand attention and concentration to complete. As Csíkszentmihályi argues, flow occurs when people:

sense that one's skills are adequate to cope with the challenges at hand, in a goal-directed, rule-bound action system that provides clear clues as to how well one is performing

(1990: 91).

Tony Butt (a surfer psychologist and ambassador for Patagonia apparel) translates the necessity for a goal-directed, rule-bound system in the context of surf-riding, stating that

> One thing that helps you reach Flow is being really clear about what you want to achieve. For example, having your mind set on perfecting a particular manoeuvre that you didn't quite pull off last time, or maybe trying some strategy for making that late take-off

(no date).

In this view, flow is attained by the human when they meet their capacity to continually hit targets they have set for themselves, and thus improve the self through practice. As Seifert and Hedderson states,

> The quality of the subjective experience is dependent upon a match between the perceived challenge and the perceived skill... enjoyment comes about as a result of the satisfaction of the need for competence through the pursuit of challenge and mastery

(2009: 110).

Flow therefore emanates from the imposition of structure and order by the isolated, rational, and self-willed human agent, on an abstract and a chaotic world. In this conception, the human is increasingly "separated" from the world, actively seeking the centrifugal tendency towards an Archimedean vantage point in which they are taken away from broader distractions in order to focus on the minutia of a manoeuvre. Indeed, for Csíkszentmihályi there would be little interest or point in undertaking any activity, including surf-riding, without the imposition of rational goal-setting and the explicit aims of self-improvement. As he describes in relation to hiking or going for a stroll:

> unless one sets goals and develops skills, walking is just featureless drudgery

(1990: 98).

The explicit attempt to attain a flow state through goal-setting and self-improvement has appeal across a spectrum of skill levels. Whatever starting position one may be in, from novice to expert, goals can be set that will extend and develop one's status, as Mark Stranger (an Australian surf scholar) acknowledges:

As a novice surfer, flow is relatively easy to achieve since a low level of skill is easily challenged and small victories, such as the first time a person successfully rides a wave, can provide thrilling experiences (2011: 166),

And as the following surf-riders state:

> I wouldn't say that I am hunting for that most awesome wave – I am just happy if I can get up and manoeuvre my board and feel like I am sort of controlling my ride on the wave if you like – I am content with that
> (Research Survey).

> The reason why I carry on surfing is just the pure challenge, you know that you can ride waves from 2 foot to 10 foot and everyone is different and also you want to get better at it – it is a sport that you are doing with nature and everything like that and you just really do want to get better at it all the time
> (Research Survey).

> Big waves are a good test – you have to commit or get a pasting! I enjoy the rush I get from sticking a new manoeuvre that I have been trying for a while or taking a big bomb or a nice deep dark pit
> (Research Survey).

Csíkszentmihályi summarises the direct causation he feels is required to attain flow in the following way:

> even the simplest physical acts become enjoyable when it is transformed [by setting targets]. The essential steps in this process are: a) to set an overall goal, and as many subgoals as are realistically feasible; b) to find ways of measuring progress in terms of the goals chosen; c) to keep concentrating on what one is doing, and to keep making finer and finer distinctions in the challenges involved in the activity; d) to develop the skills necessary to interact with the opportunities available; and e) to keep raising the stakes if the activity becomes boring
> (1990: 97).

Due to the intense focus required by the human on the skill being performed, the attainment of a flow state often means that the passage of time is sensed in different ways. As Csíkszentmihályi puts it, a "sense of time is distorted" by flow (1990: 71), and as Marsh suggests, when in the flow zone, "people engage so completely in what they are doing that they lose track of time. Hours pass in minutes" (2007: no page). The Australian surf-rider Shaun Thomson (World Champion in 1977) expresses this flow sensation as follows:

> When you [successfully] go into a deep barrel you certainly feel as if time's expanded. Life is slowed down. I felt as if I could curve that wall [of water] to my will. I really felt that. It's a magical, magical moment
> (cited in Gosch, 2008).

In a similar vein, the mental concentration required by the flow state also produces a "separation" between the surf-rider and other unnecessary or peripheral aspects of their assemblage. In Butt's words, as the surfing task demands their immediate and present involvement, their wider world will,

> reduce… right down to what you see and feel in your immediate surroundings. Nothing exists apart from you and the waves and maybe the wind or the odd seagull. Your mother-in-law, the traffic, the bank manager and the shopping have simply ceased to be
>
> (no date).

Flow is therefore a form of mindfulness activity that is task-oriented, rooted in mental concentration, and individualistically rewarding. The self is affirmed through successfully imposing a set of skill-oriented targets, and their sense of identity is reinforced when those targets are hit[3]. The notion of flow has become popular in relation to surf-riding due to the activity's basic requirement for skill competency in order to be safe and successful on the waves. The notion of flow also encourages interest in skill development, especially by those surf cultures who valorise (high) performance moves in or out of formal competition (see Part II). On the face of it, as the notion of flow focuses on and actively promotes the individualised and abstracted self, it does not resonate strongly with the relational approach emphasised in this book. However, both Csíkszentmihályi and others (see below) acknowledge the potential for this state to be extended beyond the mastery of technique and the affirmation of the individual. As the next section suggests, when the flow state is extended, this is when the relational sensibilities of surfing spaces become reoriented away from a "tendency towards separation", and towards a tendency of "mergence".

Tendency towards mergence

> If you're lucky enough to really get [through to] a total Flow situation, the whole thing will become a strange out-of the-body experience. You will feel like the whole activity is running itself and… [y]our mind and body will merge into one and you will feel like you are part of the wave
>
> (Butt, no date).

As stated above, the "tendency towards mergence" offers a different set of relational sensibilities from those associated with a tendency towards separateness (which is characterised by the flow state). When experiencing a "tendency towards mergence", the human receptor senses the various "other" components that come together to constitute the assemblage of which they are a part, and these broader cultural geographies are not experienced as distracting, distinct, or unattractive, rather they connect and extend aspects of the human in desirable ways. As a consequence, when experiencing a "tendency towards mergence", humans are not driven by a process of goal-setting and attainment, rather through their practice they become "lost" (or perhaps even found) as part of their broader relational setting. In these cases, the relational sensibilities generated can be broadly identified by the terms "convergence" and "littoral religion", and it is to these relational sensibilities that the chapter now turns.

A tendency towards convergence

> My experience is disorientating and distorted. The wave and I blend
> (Evers, 2010: 59).

The concept of convergence was initially developed in relation to post-nature (Anderson, 2009) and draws attention to how individuals may feel a strong co-constitution between humans and non-humans when encountering wild environments. The notion was later developed in with respect to surfing spaces (see Anderson, 2012), emphasising (as Evers notes above) how in the instant of surf-riding, surfers sense the thresholds of their bodies, boards, and waves becoming blurred into a unified entity/process. When a sense of convergence occurs, surf-riders do not refer to the execution of skills or the need for intense concentration (as they do when experiencing flow experiences[4]); rather they refer to the loss of a coherent sense of self as they experience the feeling of becoming part of something larger. In other words, they sense a co-ingredience with the component parts of the surfed wave. As the following surf-riders articulate:

> I love the sea and so being able to spend time in it, and *be one with sea* is fantastic
> (Research Interview, their emphasis).

> It provides a unique way of enjoying myself that is *intimately connected to nature*
> (Research Interview, my emphasis).

> [it is] the ideal of *merging with the medium* ... of a now-expiring-and-never-to-exist-on-this-planet-again miracle
> (Duane, 1996: 66, my emphasis).

For these surf-riders, being part of surfing practice is not adequately described by the successful attainment of a cut-back manoeuvre or a "hang ten" (see Chapter Five), it is better articulated as becoming subsumed by a larger set of processes, a sense of becoming "part of something that is timeless and much, much bigger than yourself" (Research Survey). As surf-scholar Scheibel notes, for those surf-riders experiencing this relational sensibility,

> there appears to be a disorientation in time and space where the surfer temporarily loses perception of all external boundaries. There is an

intensive and emotional reaction felt by the rush of adrenalin to the muscles, with the resultant feeling of emotional catharsis and the joyful sensation of having been so close in union with the ocean

(1995: 256).

Due to their affective intensity, these convergent encounters are powerful and become crystallised as significant events in the identities of surf-riders (see also Chapter Eight). The significance of relational convergence is alluded to in the following passages composed by surf writers Stephen Kotler then Fiona Capp:

As we were paddling out, the wind died, the ocean got glassy and the waves began pouring in, clean and rideable. Bigger sets started to arrive. I watched one of the [other -riders] spin his board and pick off a perfect, shoulder high, left peeler, dropping down beneath the curl, a slowly vanishing shadow sailing gleefully down the beach. I paddled fast to my left, angling toward the next wave, stroked and stood and felt the board accelerate and pumped once and into my bottom turn, and then the world vanished. There was no self, no other. For an instant, I didn't know where I ended and the wave began. ... Surfing is a game of such instants

(Kotler, 2008: 139).

Whenever I tried to pin down what this "at oneness" felt like, one particular moment in the surf always came to mind.... I remember the water swelling beneath me and how I was perfectly in tune with its rhythm. I remember a surge of energy lifting me high above the hollowing water, the thickness of the shoulder, the glowing, desert-like appearance of the shore. Above all, I remember the instant at the top of the wave, just as I rose to my feet to "take the drop", poised on the brink with the weight of the inrushing ocean behind me and the wave unfurling beneath me. The spool of my memories always froze at this last split second of clarity and separateness before the screaming descent where mind, body and wave – became one

(Capp, 2004: 11/86).

Thus convergence refers to the sense of merging or becoming one with the other assembled components of surfing spaces. Although individuals remain discrete in their material form, they nevertheless feel they have become temporarily subsumed – or converged – with other processes far larger than themselves. In these cases, surf-riding has become not a task, but has more in common with a dance in which the "partners" lose themselves within "an organic unity between man [sic] and nature" (after Flynn, in Booth, 2003: 316) [5]. Similar "tendencies towards mergence" are felt by other surf-riders when they refer to surfing spaces as spiritual in nature, and surf-riding as a "heroin injection of union" (after Kotler, 2008: 189).

Littoral religions: surf-riding as "comm-union"

As we have seen, it is possible for encounters with surfing spaces to be articulated by flow and/or convergence, and in the latter circumstance surf-riders experience relational sensibilities which point towards the blurring of (b)orders between self and assemblage. In some cases, surf-riders go further in this "tendency towards mergence" and articulate their experience in spiritual terms.

> Because every wave is different, every wave is also novel. ...Pretty soon you soon start to understand that surfers are often operating in foreign waters, far outside of their comfort zone, which, as it happens, are also the perfect conditions for producing transcendent experiences. As strange as it may be, there seems to be something in surfing that produces self-transcendent states more often than in other sports. ... To the best of my knowledge, there's no skydiving imam or NASCAR high priest
>
> (Kotler, 2008: 212/3).

As Kotler implies, being able to witness first-hand the assembling of processes in littoral locations has the potential to generate a sense of awe in surf-riders that goes beyond goal-attainment or elemental-immersion. As the following surf-riders articulate:

> the awe I felt in the presence of this natural force, a force so unfathomable that it grants you an inkling of infinity, was intrinsic to my attraction to surfing
>
> (Capp, 2004: 10).

> I enjoy being close to the power of nature
>
> (Research Survey).

> It's all about the experience of being in the sea, harnessing the power of nature to catch waves
>
> (Research Survey).

> I surf to feel at one with the immensity of nature, [it gives me both] the relaxation and adrenaline rush
>
> (Research Survey).

Due to the specific geographical components that are assembled in the littoral zone, and despite many surf-riders' expertise and experience in these locations, engaging with surfing spaces can still challenge, enthral, and humble them. Such relational sensibilities lead to many sensing something sublime in surfing spaces, tapping into notions of the transcendent and spiritual, and integrating the surf-rider beyond their worldly conceptions of time and space. As Stranger tells us,

> an appreciation of the sublime in nature came to the fore towards the end of the 18th century and involved an experience of the infinite in the contemplation of terror and beauty, as seen, in say, the Alps or stormy seas
>
> (1999: 270).

Initially explored by the eighteenth century philosopher Edmund Burke (1776), awe was identified as a relational sensibility articulated as "astonishment, mental panic and momentary amazement [that] overwhelm[ed] reason and jolt [ed] the individual into the present living moment" (Ford and Brown, 2006: 11). According to Burke, this sense of embodied awe is "the strongest emotion which the [human] is capable of feeling" (cited in Capp, 2004: 10), and many scholars have identified how this sense is often experienced in relation to nature, especially wild nature (see Oelschlager, 1991). Such cases of the sublime are often equated with sacred experiences, when a sense of energy and apprehension, mystery and fascination both attracts and repels the participant (see Otto, 1970; Graber, 1976). This sense of the sublime and the sacred in the face of nature is, in Graber's words: "A self-transcending experience which carries the mind to the edge of its limited plane of understanding" (1976: 2).

As argued by Anderson (2013), such transcendent experience is crucial to the specific sensibility of what we might describe as the "surfing-sublime". However, in contrast to many cases of the sublime, surfing is not a spectacle experienced from a distance, but it is a rather a participatory, involved encounter (although see Chapter Seven for how some spectators articulated the practice of surf-riding from the shore). Thus through becoming part of surfing spaces humans in this sense "re-enter into nature" (Fiske, 1989: 76, also cited in Ford and Brown, 2006: 17) and experience "an ecstatic union" with this assemblage (Stranger, 1999: 270). As a consequence of this relational sensibility, surfers are moved beyond the experience of the "surfing-sublime" and towards something more spiritual in nature. Surfer and writer Fiona Capp expresses this distinction as follows:

> I knew that my love of the sea was not solely derived from the attraction of the sublime, the longing for awe. Most surfers—apart from those driven purely by the competitive urge—talk openly about the sheer joy of being in the water and the visceral need for an intimate relationship with the ocean. Informing this kind of understatement is a philosophy—sometimes couched in spiritual terms—about connecting with a force vastly greater than oneself, about returning to "the source"
>
> (Capp, 2004: 10).

Although "the source" is mentioned by many surfers (see for example, Point Break, 1991, and for a more detailed discussion on the nature of "the source", see Taylor, 2008: 224), it is often impossible to define precisely what this source is and how it connects to the human spirit (see Moore, 1992: XI). Despite this, and echoing Taylor (2008), surfers do refer to the "physical, mental, and spiritual bliss" (Research Survey) that they receive from encountering the surf zone, how the "connection with creation" is "good for the soul" (Research Survey). For many surfers, therefore, "there is a massive spiritual element to surfing" (Research Interview). This spiritual element is articulated by American surf journalist Dick-Read as he reflects on comments from surf-pioneer and board-designer, Herbie Fletcher:

> "Surfing's where it's at because it takes you to this other place that everybody else wants to go to but can't get there".
>
> Hey, [these words] are not that profound, or complicated, or even grammatically correct, but Herbie Fletcher's words, which he imparts as the credits roll at the end of the excellent new surf film, Chasing the Lotus, do hit a big fat nail on the head. And he delivers them with just enough of a smirk to let you know he's kind of talking shit, but also that what's funny about this is that the shit he's talking also happens to be true
> (Dick-Read, 2007: 17).

As Fletcher outlines, engaging with the liminal space of the surf zone takes surfers to a place that "everybody... wants to go to but can't get there". This place is not just a geographical location, or simply a place of relational sensibility, but also a place of spirituality (see Moriarty and Gallagher, 2001: 10).

> Geographers should not be hesitant to recognize [surfing spaces] as a medium to understand human spirituality
> (Henderson, 1992: 472).

The worlds of nature and water have been intertwined with religious and spiritual belief in many cultures (see Sanford, 2007). Despite many scholars insisting that an appropriate definition of religion must include, "a system of beliefs and practices that are relative to superhuman beings" (Smith and Green, 1995: 893) others deny that divine beings are essential to denote practices as religious (Albanese, 2022). In sum, as a "term created by scholars for their intellectual purposes" (Smith, 1998: 281, 282), "religion" is open to diverse and subjective interpretation. As Taylor points out in a useful overview: "determining what counts as religion has become exceedingly messy" (2007b: 923).

Despite or perhaps because of this, engagement with nature and water can be understood as a form of religious ritual (see Albanese, 2022; McLuhan, 1996). Price, for example, has argued that many activities in nature can provide "a sense of wonder, awe, wholeness, harmony, ecstasy, transcendence, and solitude" (1996: 415). Moreover, he observes that the language participants use to describe their experiences in nature "frequently becomes poetic and invokes religious metaphors" (1996: 417). Taylor (2007a, 2007b, 2008) understands such "nature religion" to involve a perception that nature is sacred in some way and worthy of reverent care (Taylor, 2007b: 925), and a feeling of belonging and dependence upon the earth's living systems (Taylor, 2007b: 937, see also

Taylor, 2008). As such, many forms of environmentalism can be viewed to have a quasi-religious dimension (see Benthall, 2006; McLuhan, 1996) so too a range of outdoor activities. Rickly-Boyd, for example, has explored how practical engagement with nature through pursuits such as rock climbing may provoke existential experiences (2012), whilst Sanford (2007) has identified how involvement with water, particularly through kayaking, renders white water activity akin to religious practice. Taylor developed this idea with respect to other immersive water-based activities, including surfing; he describes such activities as "aquatic nature religions". As he states:

> "Soul surfers" consider surfing to be a profoundly meaningful practice that brings physical, psychological, and spiritual benefits .... For these individuals, surfing is a religious form in which a specific sensual practice constitutes its sacred center, and the corresponding experiences are constructed in a way that leads to a belief in nature as powerful, transformative, healing, and sacred
>
> (2007b: 923).

Even if not all surfers would define themselves as "soul surfers" (as Mitchell describes the "soul surfer is the... pinnacle of surfing spirituality, equivalent to Nirvana, Satori, Total Enlightenment, etc" (no date)), the way many surfers talk about their transformative experiences in liminal locations calls forth religious lexicon. As Taylor puts it, "it does not take long analyzing material surf culture or its associated rhetoric to see its spirituality-infused nature" (2007b: 924). As the following surfers identify:

> Life to me is a constant movement, an ever-changing, swirling mass of variables interconnecting with each other to create a whole; if you're willing to compare them to riding a wave, then comprehension of the totalness here and now is at least partially attained
>
> (Bill Hamilton, early Hawaiian surf pioneer, in Warshaw, 2005: 245).

> Surfing works in mysterious ways. It can serve as a gateway to eternal life ... Our daily baptism .... Our confessional booth ... And, of course, it can serve as our sanctuary, our church and our connection to the Supreme Being itself
>
> (Slater, 2008: no page).

With these surf-rider sensibilities in mind, the broad identification of "aquatic nature religions" can perhaps be rendered more specific in nature: these are tendencies towards mergence that specifically involve the specific littoral locations, and – as such – could legitimately be labelled a subset of this broader "aquatic" category, i.e., as "littoral religions". Indeed, magazines oriented towards the littoral are laced with religious metaphor, one of the most popular and respected, *The Surfer's Path*, alludes not only to the geographical but also to the spiritual journey surf-riders take to waves[6]. Surf writer Alex Wade, following

American surfer Tom Curren, refers to the littoral as "the blessed church of the open sky" (2004), whilst individuals often refer to their own path to and through surfing as a "pilgrimage" (see for example Anderson, 2007: 9 and Doherty, 2007). Due to the personal and unstructured nature of these "pilgrimages", surf-riders can be understood as practising a postmodern or secular spirituality, rather than a modern, institutionalised religion. Conradson (2011, 2013) provides an overview of the "messy" distinction between spirituality and religion (after Taylor, 2007b: 923) which is useful to us here. If "modern" religion tends to be organised and institutionally structured, secular spirituality "typically denotes a concern for an enduring dimension of human life that relates to issues of meaning and purpose" (Conradson, 2013: 186). As surf-riders are generally not religious pilgrims in a traditional, institutionalised sense (see Davie, 2004), they can be seen as "secular pilgrims" (Digance, 2006) practicing their own form of "cultural religion" or spirituality (Alderman, 2002). In this way, surf-riders' pilgrimages to the littoral make it possible for them, "to access God or the divine ... in [their] cosmology" (after Digance, 2003, cited in Collins-Kreiner, 2010: 440). Surfing spaces thus becomes "sacralised", with surf-riders drawn to them in order to experience something out of the ordinary, to encounter a transition from the mundane secular world of their everyday existence to a special and sacred state (Collins-Kreiner 2010: 442).

Conclusion

As we have seen, despite the range of encounters produced through the different ontologies of surfing spaces, it is possible to identify some commonalities in the ways many surf-riders articulate the "relational sensibilities" that emerge through them. In this chapter we have seen how surf-riders experience a sense of flow, or being able to be wholly mindful of their surf-riding practice and separate themselves from the non-immediate components of their (terrestrialised) assemblage. We have also identified how surf-riders experience convergence, where they sense a loss of individualised self in moments of merger within the broader components of surfing spaces. Lastly, we have seen how some surf-riders find their god in the littoral zone, when they become part of what they feel are awe-some processes at play in these spaces. Each of these relational sensibilities can be usefully framed by Seamon's notion of "encounter" and the different forms of attentive contact humans register between themselves and the assemblage of which they are a part. Whilst flow is positioned within a "tendency towards separateness", convergence and littoral religion locate humans within a "tendency towards mergence", dissolving and blending the absolute distinctions between surf-rider and surfing space. Part I now moves on to extend this discussion to the dominant lexicon with which surf-riders describe their encounters with the waves – the language of "stoke" – and how this single word has come to sum up the myriad of relational sensibilities evoked by surfing spaces.

Notes

1. As we will see in Part II, this sense of separateness may also occur in surfing spaces when "other" surf-riders or water-users interact with waves and distract "our" attention from our own engagement with the littoral location.
2. In the following sections, a range of interview and survey responses will be drawn on from the author's own primary research, these will be annotated as follows (Research Interview / Research Survey).
3. As the following respondents discuss in relation to skateboarding:

    "Wow! That is what it is like … wow! There is nothing that I have ever done in this world that has made me feel this way …. You just feel good all over, especially inside … it's an unreal feeling" [J.].

    "[Made] you feel like you have so much power and control, and no one can take this away from you. You feel so much better about yourself" [M.].

    "You get real cocky and full of yourself because you get this feeling that you can land all your tricks" [N.] (all cited in Seifert and Hedderson, 2009: 118).
4. Although this may be the required base level of their practice – in order to attain a sense of convergence they may take for granted their ability to successfully paddle, turn, and judge the correct time to catch the breaking wave. In this way, convergence may be for some the "next step" in relational sensibility from the experience of flow (as Butt implies above).
5. These moments of convergence may be significant as they also come to consolidate the incoming and outgoing iterations of practice which come to constitute the surf-riding cyborg (see Chapter Five). Becoming one with the sea, however fleetingly, may lend to surf-riders a capacity that they can sense and anticipate changes in surfing spaces, as implied by Evers in his account of his surfing tribe in Australia:

    "We know when it is cyclone season because, having grown up in the area, our bodies are alert to subtle changes in the weather. If my skin drips with sweat, there is a dreamy north-east wind blowing. This means small waves.

    My body loves the days when turbulent cyclonic winds rattle the shutters and blow through the window. I get excited and edgy. My body is tied to the weather patterns. When these winds arrive I drive to my favourite point break. Waves pound rocks and wash away sandcastles. Rain and wind swirl about. It's energising and feels like home" (2010: 47).

    The experience of convergence may also help explain some surf-riders' sense of geo-affiliation not to landed spaces, littoral locations, or the deep sea (or even particular surf moves, see below), but rather to the waves themselves. As Capp then Duane put it:

    "I could never throw off the feeling that I was only half alive when I was away from the sea. Waves loomed in my dreams" (2004: 6).

    "I didn't move to the beach to perfect my backside aerial attack (or even to learn what the hell a backside aerial attack *is*, for that matter); I moved because my need to be in the clear, alive water … on a real, honest-to-God surfboard, on a daily basis [and this] had been a source of nagging angst since the first time I'd ridden a wave" (Duane, 1996: 5).
6. *The Surfers' Path* ceased trading in 2014, reflecting the change in importance in the media architecture of surfing spaces, which is discussed in Part II.

7 The Stoke of Surf-riders

Introduction

As we have seen, despite the range of encounters produced through the different ontologies of surfing spaces, it is possible to identify some commonalities in the ways many surf-riders articulate the "relational sensibilities" that emerge through them. We have seen how surf-riders experience a sense of flow, or being able to be wholly mindful of their surf-riding practice and separate themselves from the non-immediate components of their (terrestrialised) assemblage; we have identified how surf-riders experience convergence, where they sense a loss of individualised self in moments of merger within the broader components of surfing spaces; and we have seen how some surf-riders find their source in the littoral zone, when they become part of what they feel are sublime processes at play in these spaces. Although there is no direct correspondence between the nature of assembled components in surfing spaces and the relational sensibility that emerges (for example, integrating a particular technology in a specific littoral location will not inevitably produce a singular relational sensibility), what is clear is that, for many, surfing spaces are transformative of their sense of self.

Through experiencing flow, convergence, or transcendence (perhaps even all three during one surf-riding session), many surf-riders emerge differently through their engagement with surfing spaces. In this chapter we turn to focus particularly on surf-riders' own language to describe these experiences and affects, predominantly through the relational sensibility termed "stoke". In this chapter we will briefly introduce the origins of the term "stoke", before arguing that it has developed into a hegemonic term within the cultural geographies of surfing spaces to point towards the range of relational sensibilities which affect surf-riders in these zones. The chapter will identify three reasons why this term has gained such a dominant position, firstly, due to the difficulties in articulating sensory experiences caused by what Bondi has termed the "paradox of (non-) representation" (1995); secondly, due to the globalisation of one particular surf-riding culture in the second half of the twentieth-century (see also Part II); and thirdly, the "energy-transfer" characteristics of the littoral zone which resonate strongly with the term "stoke's" meaning and affects. The chapter then discusses the range of relational

DOI: 10.4324/9781315725673-7

sensibilities evoked by the term at each stage of engagement with surfing spaces: the paddle out, the line-up, the wipe-out, and finally, the "perfect" ride.

The origins of stoke

> I was suspended in a corona of absolute bliss. Surfing, I would discover, was the subcult of stoked
>
> (Kampion and Brown, 1997: 23).

Beyond the cultural geographies of surfing spaces, the term "stoke" means to "strengthen", "fuel", "encourage", or "stir up". It has its origins in the word "stoker" – a role whose occupant "fed" or "stirred up" fires. It became associated with surf-riding cultures in California in the 1950s (Warshaw, 2005: 564), initially denoting the "enthusiasm" felt by individuals for surf-riding (see for example Ormrod, 2005). This chapter argues, however, that stoke has developed to resemble a hegemonic term for surf-riding cultures in many places across the globe, and now points towards a range of often ineffable, difficult to describe, and perhaps even pre-literate relational sensibilities which emerge through engagement with surfing spaces. By turns, therefore, stoke may refer to the hedonistic thrill generated by surf-riding or the fear felt in the face of large waves; it may point towards a sense of the transcendent, a moment of merger, or a personal achievement secured. Stoke as a single word has evolved, therefore, to sum up the myriad relational sensibilities evoked by surfing spaces.

Stoke as a many and varied relational sensibility

In the most popular sense, as Fordham states, stoke refers to "the feeling of euphoria stimulated by surfing" (2008: 265), and as Evers notes, "if one is very stoked, [you] experience a fully embodied feeling of satisfaction, joy, and pride. You will tingle from your head to your toes" (2006: 230). Yet stoke is also used to describe many other aspects of the surf-riding experience. Stoke may arise through anticipation for waves (as Ford and Brown cite, surf writer Craig Coombs feels this strongly when he looks at surf photography: "The picture on this page gets me so amped for a barrel I can't even sit still. Just looking at it makes me want to jump up and run around yelling" (2006: 125)). Stoke can be used to refer to the complex connection with flow and fear, as legendary big-wave surfer Bruce Irons details on a near-death experience:

> that wave at Teahupoo [Hawai'i]... was probably the heaviest wave I've ever been on. I fell off backwards upside-down, and remember saying to myself, "This is the worst situation I have ever been in surfing" ...but I was stoked, thinking, "I just went through the heaviest trip ever and I pulled it"
>
> (2012: no page).

Stoke may also refer to the satisfaction felt when surf-riders reminisce about their practice, as US surf-rider DC Green outlines, "If not for regular adrenalin hits from surfing, I'd probably own multiple suits, none of which would bring me any joy….When I'm decaying in an old folks' home, I know I'll be stoked at a life well lived" (2007: 48). Stoke may be used to describe the drug-like affect some experience through surf-riding practice, as Casey notes, stoke is:

> like a collage of sensory impressions. There may be a flash of white spray, a sudden jolt, a feeling of energy surging beneath your feet, the suspension of time so that ten seconds stretch like taffy across a violent blue universe. Inside the barrel, a place that surfers regard with reverence, light and water and motion add up to something transcendent. It's an exquisite suspension of all things mundane, in which nothing matters but living in that particular instant. Some people spend thirty years meditating to capture this feeling. Others ingest psychedelic drugs. For big-wave surfers, a brief ride on a mountain of water does the trick
>
> (2011: 75).

Indeed, stoke may be used to sum up the transcendent, as Lovett states:

> Stoke is like a tap root that connects us to the amniotic fluid of consciousness in Goswami's Ground of Being, to the quantum state of mind where the surfer, the board, hand glide or body, together with the wave and the ocean, momentarily all dissolve into one
>
> (2015: 236).

Stoke can be used to refer to all these surf-riding experiences and more, it is, according to Kampion and Brown, "one of those wordless conditions that you have to experience to know" (2003: 200). Stoke is therefore a simple word which alludes to both a straightforward and intensely complex set of relational sensibilities, stoke is like "being unplugged from life for a second. God, it's the neatest feeling" (Edwards, cited in Stecyk, 2002: 48).

Stoke: A working solution to the paradox of representation

The relational sensibilities that emerge in the surf-rider cyborg are therefore manifold and complex. As Dick-Read, when editor of *The Surfers' Path*, states in reference to surf-riding encounters:

> … Everyone [knows surf-riding is] cool… [We] know there's magic in it. But …we can't explain it. …Although we *do* it and know it we can't actually *nail* it verbally either
>
> (2007: 17).

Stoke can therefore be seen to be a working solution to a problem that faces all humans, including surf-riding cyborgs: what Bondi terms the "paradox of (non-)representation" (1995). This paradox refers to the situation that although humans experience phenomena, when they wish to reflect on or share these experiences, the range of vocabularies they have access to can never fully *be* those encounters, they can never fully "convert, translate and represent" their experiences to others (Ford and Brown, 2006: 174). Thus even though surfers know first-hand the range of affects that emerge when they are co-constituted with a surfing space assemblage, they are nevertheless unable to communicate these complex sensibilities to others in a way that guarantees coherent, reciprocal understanding. As surf journalist Medeiros points out,

> wave-riding experience[s] lie beyond language. There is an invisible and yet insurmountable wall separating the symbolic world and the act of surfing… which brings us to this horrifying and ultimately mind-fuck of a conclusion:… At best, [surfing] can only be felt – but never, ever shared or represented
>
> (2010: 95).

In the face of the paradox of representation, the term stoke provides a working solution. When surfers' experiences escape their ability to pin them down precisely, this relatively open term functions as a broad representational container into which a range of relational sensibilities can be poured, then shared. Adopting this vernacular term establishes for surfers their own linguistic cultural space in which stories can be communicated, and experiences can be connected – what Thrift (in a different context) describes as an "intersubjective space of common action" (1996: 39). Stoke, to some extent at least, overcomes surfers' inability to precisely communicate, or receive, wholly accurate re-presentations of the particular assemblages of which they were a part.

Stoke as cultural and energy transfer

As we have seen, the term "stoke" has the capacity to function as an "intersubjective space of common action" and understanding, and this goes some way to explain its popularity and near-universal adoption in surfing spaces. Yet the globalisation of the term is also directly connected to the influence of what Thomson (2017) describes as the "Californication" of surf riding since the 1950s (see Part II). As Warshaw suggests (above), the term stoke was adopted as slang by surf-riders in California, and as their own version of surfing space culture was exported and translocated in the decades that followed, the term itself has travelled with it. As it has done so, mimicking perhaps phrases such as "okay" and gestures such as the "shaka", "stoke" has morphed into a simple, easy-to-recognise "global surf-rider" vocabulary (see Part II) which connects surf-riders together regardless of their geographical or cultural origin.

Despite the relatively open nature of the term, the durability and persistence of the word "stoke" can also be traced to its resonance with the characteristics of the littoral location. As we have seen in Chapter 3, littoral locations can be understood as places of energy transfer. When surf-riders surf, they are not riding water as such, rather they are moving on the surface of kinetic energy which is moving through this media. Surfing spaces are worlds of energy, spaces for energy release, interactional transfer, and generative affects/effects. It can be argued, therefore, that surf-riders are re-fuelled and stirred up – they are "stoked" – through becoming part of this world of energy. It is as if the energy that is transferred from the sun, through winds, to the surface of the sea, is transferred through the breaking wave into the surf-rider themselves. In this way, surfing spaces literally stoke their riders. Despite the paradox of representation, stoke nevertheless has the effect of conjuring up the range of generative capacities of this "energy transfer" to the surf-rider, as the following writers re-present:

> The only place he had to call his own – outside of the front seat of his truck – was the arching face of an incoming wave. As the narrator in his brain stated: The ephemeral nature of these substantially imaginative encounters with waves approaches the illusory, yet the real-time neural stimulation and restorative prophylaxis suggests concrete uploading of significant bioremedial components, resulting in a stunning net profit on time and energy invested
> (Kampion, 2004: 44).

> Surf journalist Mike McGinty described in 1993 how difficult the surf-writing task can be. 'I don't need paper and ink, I need 24-karat gold monster-cable speaker wire with one end plugged into Backdoor Pipeline and the other soldered into your adrenal gland'
> (Warshaw, 2004: XX).

> The power under my feet was a rare sensation – that bliss of feeling the perfect line on a wave that is stacking up exactly as you want it to…. I paddled back out, *energised* by the feeling I'd just had – the wind in my hair as I raced across the wave, the responding board under my feet, water drawing up the wave face, *transferring its power* through my body.
> (Anderson, 2010: 40, my emphases).

> Wave approaching, paddle hard toward its steepest section, turn, and paddle back with it. Feeling the board gliding on its own, hop in one motion from prone to upright, slip down the face, lean to the right, turn up into the wall and laugh out loud at the watery road rising up to greet us, step a little forward to speed across the steep spots, drag a finger in the water just to believe it's really happening, and feel the light joy of effortless, combustion-free speed. And then, in the moment of detumescence, flop off the board *with just a little more juice than when you started*
> (Duane, 1996:11, my emphasis)[1].

94 *The Stoke of Surf-riders*

In the remaining sections of this chapter we explore the ways in which surf-riders have attempted to communicate the relational sensibilities of this energy transfer within the surfing space assemblage, in other words, how they attempt to articulate the myriad facets of "stoke". This exploration will occur from a number of perspectives: firstly from the "outside", from the point of view of a non-surfer watching surf-riding from a distance; secondly, from "out back", through the waves and waiting for the ride; then, the "wipeout", from the perspective of being caught inside a breaking wave; and finally, the feeling of a "perfect ride".

Surf-riding spectacle

First of all we consider the spectacle and spectator view of surf-riding. This outsider perspective is generally considered to be the most mediated and removed from the stoke of surfing spaces, but in the following contemporary and historical excerpts, these outsiders point towards the stoke they perceive emerging in surf-riders from the shore.

> In one place we came upon a large company of naked natives [sic], of both sexes and all ages, amusing themselves with the national pastime of surf-bathing. Each heathen [sic] would paddle three or four hundred yards out to sea (taking a short board with him [sic]), then face the shore and wait for a particularly prodigious billow to come along; at the right moment he would fling his board upon its foamy crest and himself upon the board, and here he would come whizzing by like a bombshell! It did not seem that a lightning express train could shoot along at a more hair-lifting speed
>
> (Mark Twain, 2004: 8).

> He went out from shore till he was near the place where the swell begins to take its rise; and, watching its first motion very attentively, paddled before it with great quickness, till he found that it overlooked him, and had acquired sufficient force to carry his canoe before it without passing underneath. He then sat motionless, and was carried along at the same swift rate as the wave, till it landed him upon the beach. Then he started out, emptied his canoe, and went in search of another swell. I could not help concluding that this man felt the most supreme pleasure while he was driven on so fast and so smoothly by the sea...
>
> (Captain James Cook, 2004: 4).

> September evenings they are here after work,
> The light banished from the sky behind,
> An industrial sunset oiling the sea.
> I watch them emerge from the last wave,
> Young men and girls grinning like dolphins

In their rubbers, surf-riders swept
Suddenly onto this table of dark sand
And thrift, the coastline's low moraine.
...
Theirs, briefly, is a perilous excitement
When the current lifts them high
And they stand erect on roofs of water,
Balanced on the summit of a wave.
...
And they creep to shore exhausted,
Barefoot, wincing with the discriminate
Steps of thieves, aware perhaps
Of something they might have won, or stolen
                (Minhinnick, 1999).

Perhaps the most noted shore-side description of stoke being generated through surf-riding is penned by Jack London (which as we will see in Part II, was central to the commodification of Hawai'ian surfing for white American by entrepreneurs like Alexander Hume Ford):

> The grass grows right down to the water at Waikiki Beach, and within fifty feet of the everlasting sea. The trees also grow down to the salty edge of things, and one sits in their shade and looks seaward at a majestic surf thundering in on the beach to one's very feet. Half a mile out, where is the reef, the white-headed combers thrust suddenly skyward out of the placid turquoise-blue and come rolling in to shore. One after another they come, a mile long, with smoking crests, the white battalions of the infinite army of the sea. And one sits and listens to the perpetual roar, and watches the unending procession, and feels tiny and fragile before this tremendous force expressing itself in fury and foam and sound. Indeed, one feels microscopically small, and the thought that one may wrestle with this sea raises in one's imagination a thrill of apprehension, almost of fear. Why, they are a mile long, these bull-mouthed monsters, and they weigh a thousand tons, and they charge in to shore faster than a man can run. What chance? No chance at all, is the verdict of the shrinking ego; and one sits, and looks, and listens, and thinks the grass and the shade are a pretty good place in which to be.
> And suddenly, out there where a big smoker lifts skyward, rising like a sea-god from out of the welter of spume and churning white, on the giddy, toppling, overhanging and downfalling, precarious crest appears the dark head of a man. Swiftly he rises through the rushing white. His black shoulders, his chest, his loins, his limbs—all is abruptly projected on one's vision. Where but the moment before was only the wide desolation and invincible roar, is now a man, erect, full-statured, not struggling frantically in that wild movement, not buried and crushed and buffeted by those

mighty monsters, but standing above them all, calm and superb, poised on the giddy summit, his feet buried in the churning foam, the salt smoke rising to his knees, and all the rest of him in the free air and flashing sunlight, and he is flying through the air, flying forward, flying fast as the surge on which he stands. He is a Mercury – a brown Mercury. His heels are winged, and in them is the swiftness of the sea. In truth, from out of the sea he has leaped upon the back of the sea, and he is riding the sea that roars and bellows and cannot shake him from its back. But no frantic outreaching and balancing is his. He is impassive, motionless as a statue carved suddenly by some miracle out of the sea's depth from which he rose. And straight on toward shore he flies on his winged heels and the white crest of the breaker. There is a wild burst of foam, a long tumultuous rushing sound as the breaker falls futile and spent on the beach at your feet; and there, at your feet steps calmly ashore a Kanaka, burnt golden and brown by the tropic sun. ... He is a... member of the kingly species that has mastered matter and the brutes and lorded it over creation

(London, 1911).

From outside to outback

From his initial description of the "winged Mercury" (from which the surf corporation Quiksilver takes its name), Jack London attempted the practice of surf-riding himself. In this process of going from "outside" to "outback" London usefully recounts the nature of wave formation discussed fully in Chapter Three:

And that is how it came about that I tackled surf-riding. And now that I have tackled it, more than ever do I hold it to be a royal sport. ...Get out on a flat board, six feet long, two feet wide, and roughly oval in shape. Lie down upon it like a small boy on a coaster and paddle with your hands out to deep water, where the waves begin to crest. Lie out there quietly on the board. Sea after sea breaks before, behind, and under and over you, and rushes in to shore, leaving you behind. When a wave crests, it gets steeper. Imagine yourself, on your board, on the face of that steep slope. If it stood still, you would slide down just as a boy slides down a hill on his coaster. "But," you object, "the wave doesn't stand still." Very true, but the water composing the wave stands still, and there you have the secret. If ever you start sliding down the face of that wave, you'll keep on sliding and you'll never reach the bottom. Please don't laugh. The face of that wave may be only six feet, yet you can slide down it a quarter of a mile, or half a mile, and not reach the bottom. For, see, since a wave is only a communicated agitation or impetus, and since the water that composes a wave is changing every instant, new water is rising into the wave as fast as the wave travels. You slide down this new water, and yet remain in your old position on the wave, sliding down the still newer water that is rising and forming the

> wave. You slide precisely as fast as the wave travels..., th[e] water obligingly heaps itself into the wave, gravity does the rest, and down you go, sliding the whole length of it.... I shall never forget the first big wave I caught out there in the deep water. I saw it coming, turned my back on it and paddled for dear life. Faster and faster my board went, till it seemed my arms would drop off. What was happening behind me I could not tell.... I heard the crest of the wave hissing and churning, and then my board was lifted and flung forward. I scarcely knew what happened the first half-minute. Though I kept my eyes open, I could not see anything, for I was buried in the rushing white of the crest
>
> (London, 1911).

The out-back perspective of surf-riding practice thus offers a further insight in the nature of the activity, one that is unobservable from the shore; as the South Walian surfer "Fuz" identifies:

> ...when you're out the back, sitting in the line up on a strong offshore day, and you can see rainbows on the backs of breaking waves. You take off, drop down the face, the spray is blowing back at you and again, if you look round, there's a brief, flickering rainbow. People on the shore just can't see this, they can't experience this feeling... For a surfer, the fluidity of the ocean – of being in it – is what matters
>
> (in Wade, 2007: 63).

Fear and danger: Being caught inside

Alongside the thrill that the term stoke involves, the risks that are intrinsic to surf-riding (see Chapter One) also means that feelings of fear and threat are inherent within the term. This vital but lethal aspect of stoke is articulated by American surfer Greg Page as follows:

> It was only a matter of moments before the biggest wave I had ever come face to face with finally arrived. I wasn't going to let it go without me. I spun my board around, put my nose down, and paddled as hard as I could. I felt the wave lift behind me, and even though I was paddling like crazy, I was still being lifted up, almost going backwards. I was looking down an impossible steep wave that was two storeys tall when I felt the sudden acceleration and I stood up. I zoomed down the face of th[e] wave faster than I had ever gone before, but not fast enough.
>
> I had taken the first wave of the set and I was about to pay the price for not choosing more carefully. I made it to the bottom just before it closed out. I straightened out in front of the mountain of whitewater following me and lay on my board, hoping to prone it out. The impact hit me like a truck. I went spinning and tumbling like a rag doll. I punched myself in the nose as I cartwheeled in the cauldron, and then I was pinned on the

bottom in the cold darkness as the violence rumbled by me. I was craving air when I was finally able to push myself off the bottom and head for daylight.

The surface was thick with foam, and just before I got my head up for breath, the next wave hit me. My lungs were aching as I was driven down and over and around and around. I was lost. Didn't know which way was up. Needed a breath. Seeing stars. Relax. Panic. No, relax. Getting dark. My feet found the bottom, and I gave one last push and broke the surface gulping air and foam

(2003: 311–12).

Clifton Evers describes the relational sensibilities of a wipe out in the following way:

I spin, paddle, push myself over the edge, leap to my feet and slide. The adrenaline has gotten to me and I feel invincible. The trim of the surfboard is high and fast, and I angle towards the bottom of the wave. My surfboard bounces but my legs absorb the shock. With feet spread wide I skirt near the pitching leap. Too near. Pow!

I fall before I can prepare and my head slaps the water hard, my skin goes tight in shock, and everything looks yellow and purple. Surprised and startled by the fear of disunion, my assumed governance of my body slips away…

There is a lack of control as I tumble, twist, contort and merge with bubbles. It's a thorough thrashing. Clouds of air explode and water sweeps me into a cave pocketing the reef. It's quiet and dark in the depths and hard to work out which way is up. There is a sense of abandonment and I am forced to just "go with it". My body floats upwards and I use the sensation to work out where to swim. My senses, though, are working overtime to escape the turbulence threatening to smear me onto the cliffs.

One side of my body is numb and my lip is split from the pressure of the implosion. Blood leaks into the sea water, with which it is more or less interchangeable. My board is silhouetted above me. Breaking the surface I gasp for air. The next wave hits and water travels down my throat. I convulse. Down, down, down I go. Then, without warning, the ocean lets me go. I get a tiny breath of salty foam and air, briefly go under again, and then resurface. A mate paddles over and I think he asks if I'm OK. I am dazed and confused. I nod my head. A sneaker set appears. Fear flags the present and snaps me back into the here and now. I look down into the trough as I travel over a large wave to see a cascading waterfall smash another surfer as her hands grasp at the water

(2010: 59).

Facing the fear generated through engaging with surfing spaces is a necessary dimension of surf-riding. Due to the inevitability of risk and danger, and the subsequent anticipation of your reaction to it, waves become "as much in the mind as in the sea" (Capp, 2004: 129). Alan Weizbecker, an American surf writer, captures this inevitable confrontation with your own emotions in the following piece:

> ... The previous afternoon, while cleaning and patching my cuts and extracting urchin spines, I'd decided to take a day or two off from surfing to let my wounds heal. But now I was remembering certain waves I'd locked into over the past three days... A fluttering in my stomach indicated what I was thinking about thinking about before I actually thought it: I was thinking about thinking about going back out.
>
> I've been told that skydiving is scarier the second time than the first, the reason being that the second time you know from experience what a ridiculous, stupid act jumping out of an airplane really is. It's not just ridiculous and stupid *in theory*. You've done it, so you *know*. And you also know just how fearful you're going to be when you get up there and you're in the open door looking down; your fear is now a fear of fear
>
> (2001: 46).

Facing fear for the "perfect ride"

Overcoming the fear of fear enables further engagement with surfing spaces, and the possibility of encountering your "perfect ride". As we know from Part I thus far, what the "perfect ride" may be is not a homogenous experience or assemblage, and will be dependent upon the individual surf-rider, their skill, and their technology. For our purposes here, however, it is worthwhile noting some core aspects of what may help to constitute a "perfect ride" for many surf-riding cyborgs. The perfect ride may perhaps occur on a breaking wave which has a strong swell and long fetch, it will be formed in conjunction with a steeply sloping foreshore and light offshore wind, and these components will combine to generate a degree of vertical mobility in the wave, causing it to rise, peel, and curl, and perhaps even create a tube through which, if positioned correctly with suitable speed, technology, and skill, the surf-rider can travel. Ideally, this ride will be earned through vernacular understanding of the combined processes at play, it will reward patience, and commitment, and will not be interrupted by other water presences. For some, this ride should be witnessed by fellow surf-riders in order to share the experience vicariously and enhance it through post-event camaraderie. A version of the perfect ride is well articulated in the following passage by British surf writer, Tom Anderson:

> After about five attempts to get in and around the tube, one virtually landed on my lap. As if meant for me, everything suddenly slipped into place and all semblances of things being hard, difficult, tricky or frightening took ten seconds' leave. Just enough to allow me a ride that would stay with me forever. A peak so round and solid that you wanted to cry with excitement made its way through the ocean towards me, just as a burst of

rain began to pour out of the light-grey skies. Apart from where the droplets broke its surface, the water appeared silvery, slippery, thick. There wasn't a wisp of wind and I was in exactly the right position to go for it, with no hassle from anyone else.

I paddled, coolly focusing my breathing and senses to channel the excitement in my muscles into the economy of motion needed to negotiate the drop. Building up paddle momentum I got over the ledge well in time to avoid any kind of freefall, but still deep enough along the reef to find a long wall of mercury-like water rising up in front of me. Way beyond, out in the flats and a world I had left behind, I could see [fellow surf-rider] Rhino sitting up and raising his hands into the sky. ...Time began slowing and my awareness of all around me heightened to a haywire crescendo. I could hear my breathing as my thought processes clarified... *This wave is going to do it... It's going to do it... It's going to...*

The lip hooked itself outwards, piercing the flat water to my left, swallowing me in the back of its saltwater pocket. With so much room, my board was able to stick to a clean wave face and continue unhindered in its trajectory forward, towards the window of light that had now shrunk my view of the channel to merely a snapshot – with Rhino cheering ecstatically in the middle of it. This time the absolute change of sound, gravity and atmosphere indicated how far behind the portal I was. There was time to think, to stare, to marvel – then as quickly and predictably as it had thrown over, the exit suddenly flew towards me and I was catapulted out and back into reality, careering onto the shoulder with runaway speed and a grin that could be seen from the town centre. Taking a breather to register what had happened, I let the euphoria flood through me and waited for my psyche to adjust. I had to try and do it again

(2010: 78).

Despite the "perfect ride" and "perfect wave" becoming broadly scripted in line with that illustrated above (see also Part II), as stated above it remains important to emphasise that due to the range of ontologies that constitute surfing spaces, there is no one version of perfection that holds true in all times and in all littoral locations. The perfect ride and perfect wave are therefore poly-versal rather than monolithic in nature, overspilling narrow definitions and varying in form, context, and size depending on the goals of the aspiring surf-rider, their skills, technologies, and execution. As a consequence, and combined with the multitude of disparate components which must come together in a moment to realise surfing spaces, it may be more useful to conceive of the perfect ride as *any* which is fortunate enough to occur; in this sense, "stoke" is the experiential outcome registered in the surf-rider who is lucky enough to be part of the coming together of weather, water, geology and technology which constitutes a surfed wave.

Conclusion

In this chapter we have explored the ways in which many surf-riding cyborgs attempt to articulate the range of relational sensibilities that emerge through their co-constitution with the breaking wave. We have seen how the deceptively simple word "stoke" can function as both a noun and a verb, and has become widely adopted in order to point towards this range of affects which "fuel", "encourage", and "stir up" the surf-riding cyborg. Acknowledging the word stoke is significant when understanding surfing spaces as it reminds us of the importance of re-presenting the surf-riding experience outside of that experience – whilst replaying and processing the encounter for yourself and for others is a vital aspect of the human condition, it also presses us against the limits of re-presentation as we do so. Stoke therefore functions as a term through which a common space of understanding can be generated, and the embodied knowledges of surf-riding can at least be gestured towards, and subsequently acknowledged and affirmed by others. The term's popularity also draws our attention to ways in which languages from one regional surfing location (i.e. California) can gain traction more widely (a point we will return to more extensively in Part II), and how the experience of "stoke" can be transformative to how surf-riders see themselves, and the world. It is this transformative affect that the final chapter in Part I develops further, and we now turn to critically explore the notion of the surfing space assemblage as an "event".

Note

1 The dominance of male, white voices in specific surfing spaces, and their potentially "toxic" consequences, are being increasingly noted by both surf-riders and -scholars (see Part II), yet on occasions such voices do capture key dimensions of the issues circulating round stoke as representation. The following excerpt from Australian surfer and writer Tim Winton's novel *Breath* not only suitably articulates the problems that surf-riders face in putting relational sensibilities into words, it also implies the influence of particularly masculine cultures in shaping surf-riding more broadly, whilst alluding to how experiencing surfing spaces can transfer-energy to its constitutive components (including the surf-rider). As Winton writes:

"It was hard to find the words for the things we'd just seen and done. The events themselves resonated in our limbs. You felt shot full and the sensation burned for hours – for days, sometimes – yet you couldn't make it real for anybody else. You couldn't and you weren't sure you wanted to. But we blathered at each other from sheer excitement and you can imagine the boyish superlatives and the jargon we employed.

[...Yet] Sando was good at portraying the moment you found yourself at your limit, when things multiplied around you like an hallucination. He could describe the weird, reptilian thing that happened to you: the cold, supercharged certainty which overtook your usually dithering mind, the rest of the world in a slow-motion blur around you, the tunnel vision, the surrender that confidence finally became. And when he talked about the final rush, the sense of release you felt at the end, skittering out to safety in the beautiful deep channel... "It's like you come pouring back into yourself", said Sando one afternoon. 'Like you've exploded and all the pieces of you are reassembling themselves. You're new. Shimmering. Alive'

'Yes', she said. 'Exactly'" (Winton, 2008: 128).

8 The Event of Surfing Spaces

Introduction

We have seen in Part I how a range of physical processes come together in littoral locations to create space in which surf occurs; we have also examined how, through a range of performative encounters, humans and their chosen technologies intercept these processes and emerge as part of surfing spaces. These chapters have argued that surfing space assemblages are generated through encounters which are transitory and fleeting, yet despite their ephemeral nature have the capacity to be transformative: through their coming together they change the capabilities and properties of the components that are temporarily part of their composition.

As Brubaker suggests, it is crucial to do two key things when engaging with such ephemeral but significant spatial encounters; it is important to "give serious theoretical attention to [(firstly) the composition of] contingent events [and (secondly)] their transformative consequences" (1996: 21). The chapters in Part I thus far have responded to Brubaker's first objective. They have sought to locate the assemblage of surfing spaces in the littoral zone and examine how surf-riders emerge dispersed and extended through practical engagements with this space. They have also demonstrated how a range of relational sensibilities are generated in surf-riders which register the energy transferred from the physical components of surfing spaces to the human. The remaining chapter in Part I theoretically gathers the insights from those preceding it in order to begin addressing Brubaker's second objective: the transformative consequences of these spatial encounters. Chapter Eight argues that the embodied, affective, and relational consequences of becoming part of surfing spaces can be understood by employing Badiou's conception of the "event" (2001, 2005, 2006).

In this chapter surfing spaces will be argued to be "eventful" in nature. In lay vocabulary, an "event" is simply something "that happens". However, for French theorist Alain Badiou, it is important to emphasise that any event has the potential to realise a *sustained transformation* of the world. As a consequence, Badiou has refined the common understanding of the event to refer to not just any spatial moment, but one that realises its transformative potential – in other words, as one which brings something significant, persistent, and "*new* to the

DOI: 10.4324/9781315725673-8

world" (after Bassett, 2008: 895, my emphasis). A Badiourian event is therefore an occurrence that realises its innate potential to offer the exceptional, bringing about a "moment of *rupture* in time and space" (ibid) and enduringly transforms the normal conditions under which life is lived. To this end, there are four interrelated aspects to a Badiourian event. Firstly, there must be an "irruptive" moment which brings something new to those experiencing it; secondly, this moment must offer transformative potential; thirdly, "fidelity" must be shown to the irruptive moment; and fourthly, the potential for transformation must be realised in practice. Due to the relative rarity of such phenomena, Badiourian events are commonly considered to occur on the scale of the dramatic, historical, and international (and include oft-cited exemplars of the Paris Commune and Christ's resurrection; see Bassett, 2008). However, as we shall see, it is also possible for Badiourian events to occur on the scale of the intimate and the local[1], indeed this chapter suggests that surfing spaces can be understood as Badiourian events. It posits that when humans emerge through their contingent engagement with surfing spaces, new relational sensibilities are generated which are irruptive and significant (such as those articulated in Chapters Six and Seven). These experiences are transformative in nature as they re-form surf-riders' sense of self, and, as a consequence, cause them to re-orient their aspirations, activities, and circumstances in line with the implications of their experience. Surf-riders' fidelity to the subject-forming experience of the event prompts the range of cultural practices which will be examined in Part II, *Scripting Surfing Spaces*.

Badiourian events

Firstly, for Badiou, an event involves an irruptive moment. This moment, as Bassett has identified, introduces something new to those experiencing the occurrence, as Nancy explains:

> What makes the event an event is not only that it happens, but that it surprises – and maybe even that it surprises itself...
>
> (Nancy, 2000: 159).

Thus beyond the straightforward definition that an event is something that happens, for Badiou an event is also an occurrence that cannot be predicted, constructed, or planned for; it is an "unforeseen happening" (Barker, 2002: 6). The event itself thus surprises those who experience it, and as a consequence, breaches the prevailing norms and orthodoxies of their life. As Bassett identifies:

> Events are ruptures within the established order of things, which happen in certain times and places in unpredictable ways. They typically begin in confusion and obscurity with no assignable cause, and they cannot be inferred from the situation within which they erupt
>
> (2008: 898).

Badiourian events are therefore unpredictable in nature, they are beyond or before the mundane. Their capacity to surprise offers the potential of magic and enchantment, threat or fright. As Laclau states, an event is so irruptive of the everyday that it is extraordinary; "its break is truly foundational" (2004: 121).

Although events can be fleeting moments, the irruptive nature of these moments means their second key feature is their transformative effect. This effect is difficult to escape, for as Zizek puts it, "When something truly New emerges, one cannot go on as if it did not happen, since the very face of this innovation changes all the coordinates" of normality (2004: 175). For those affected by it, an event is therefore likely to change the way in which the world is viewed, and how one relates to it. However, this transformative effect can only be judged as such if, over time, significant change can be seen to have occurred as a consequence. As Latour notes in relation to the example of falling in love:

> The veracity of truth [that an event is indeed an event] is not based here on fact but rather on felicity, of the experience of believing, *and then the fidelity that is the subsequent action etched out of this experience*
> (Latour, 2005: 230, also cited in Dewsbury, 2007: 453, my emphasis).

In this example, the transformative potential of falling in love is only realised through the subsequent fidelity of individual to another (for example) which renders love, and the act of falling in love, meaningful. For Badiou, it is this fidelity to the event that is the third distinguishing feature of the phenomenon. The transformative a/ effect of the event is thus only realised when individuals are moved to demonstrate loyalty and commitment to it, in other words, the event prompts "fidelity". Only if commitment is promised, given, and maintained, then the event realises itself. As Dewsbury states, "moments thus only become events in their maintenance through fidelity post their taking place" (2007: 453). In short, an event is marked out from a mundane moment or everyday situation by this fidelity (see Barker, 2002).

Demonstrating loyalty to the implications of the event therefore defines the event as such – and defines the person experiencing it. Thus fidelity is both event- and subject-forming, and this, according to Badiou, is the event's final distinguishing feature. As Hallward explains: "some human beings *become* subjects [of an event]; those who act in fidelity to a chance encounter with an event which disrupts the situation they find themselves in" (2004: 2). For these individuals, the event changes their life, and they re-orient their aspirations, activities, and circumstances in order to constructively align with its experiences and implications. These four interrelated components: the irruptive moment, the transformative potential, the fidelity shown, and the subject-forming aftermath, define the event in a Badiourian sense.

Surfing Spaces as eventful

It is possible to understand surfing spaces as eventful in nature. As we have seen, the first distinguishing feature of an event is that it is an "irruptive moment", it "ruptures" prevailing norms and orthodoxies for those who experience it. As we have seen in preceding chapters, for those that encounter them, surfing spaces can create novel and extraordinary relational sensibilities which they cannot experience in any other way. As a moment that is by turns surprising, spectacular, sublime, and scary, the feelings of being co-constituted by the coincidence of swell, fetch, tide, craft, and body is unlike anything experienced by the "individual" up to that point. As Mike Hynson, the star of the seminal surf movie, *Endless Summer* (and for more on this film, see Part II), states in an interview with *Drift* magazine:

MH: What's really funny is that I didn't start surfing until I moved to California... So I was about 17, it became something that everyone started doing. I went down to the beach one summer and kinda went "jeez, how did I miss this?"
DRIFT: You got the bug early on?
MH: Oh my god yeah – from the very first wave! I was stood up and looking in, and I realised I had my balance and these doors either side of you just open up. It's a funny experience; you're weightless

(Mojo, 2008: 62).

And as American surfer Greg Page puts it:

I'll never forget my first wave. My Dad pried me up from my death grip to a tentative stand-up position on the front of his ten-foot board. I've been a surfer ever since chasing waves, being chased by sharks, and chasing the dollar so I can surf some more

(Page, 2003: 307).

In Page's comment, the irruptive moment of his first wave is immediately connected to how his identity was changed as a consequence of this experience, as well as referring to his ongoing commitment to surfing space encounters. Indeed, the irruptive charge of the first ride has become a common tale within surfing spaces, and the resonant understanding (see Anderson, 2002) of the eventful nature of this experience is encapsulated in the following example shared by American surf-writer Duane:

Nearby, a boy no more than eight years old, suit bagging around him as he lay on a huge longboard, took a shove into a wave from his dad and made it successfully to his feet. Suddenly, he found himself zipping sideways through the sunshine, and the shock almost overwhelmed him. He screamed in his shrill little voice, "No way! Wow! Oh, man!" with such an unbridled joy – so out of the code of taciturn surfer cool – that every man in the water, tough guys included, smiled magnanimously.
"Well, that's that," said a portly guy on a huge board to the proud father. "You can forget about him ever being President"

(Duane, 1996: 13).

The irruptive moment of engagement with surfing spaces is most commonly experienced first-hand, but (reflecting Chapter Seven) can sometimes be experienced vicariously. An example of this is the following tale articulated by (then novice) UK surfer Tom Anderson who narrates the irruptive moment of witnessing a "perfect wave" in South Africa on a television screen in South Wales pub, and American surf-rider Tom Curren catching the "perfect ride" within it[2]:

> I'll never forget sitting there, still only a boy, watching Curren's first ride at J-Bay on the big screen in the function room of a Porthcawl pub. All the best local guys had made it down there that night, to watch the wave of the decade. It'll stay with me forever, seeing some of the top surfers in my area (they surfed as well as I imagined it was ever possible to) just losing it as they watched the first under-the-lip fade, gasping as Curren emerged from tube number two, and then screaming and erupting into applause as he dropped out of sight and into the section they call "Impossibles". It was fantasy… Hard to imagine that this "wave", this mass of water so suited to surfing, actually existed; and that Tom Curren was a real human being who had gone there and taken off on it
>
> (2007: 13).

Becoming part of surfing spaces can therefore be considered as an irruptive moment (or filled with irruptive moments, as the experience of watching new waves in new places demonstrates in Anderson's account above). In line with Badiou's account, these "moments of rupture" bring "something new" to the surf-riders' world which have the potential to be transformative. Whether experienced vicariously or otherwise, these moments conjure up a relational sensibility for those co-constituted by it. As Anderson narrates, this feeling, prompted through their temporary but irruptive convergence with the assemblage of the wave, is unlike anything they have experienced before – it is both unpredictable and enchanting, so irruptive that (literally in the case of the J-Bay tube) "its break is truly foundational" (Laclau, 2004: 121). Surf-writer Susan Casey echoes these sentiments when she considers her own experience of watching big waves:

> I wondered …why it was [that], after eighteen years, I couldn't stop thinking about that day at Sunset Beach [off the North Shore, Hawai'i]. Far from being an abstraction in the ether – like electrical waves, X-ray waves, or radio waves – those thirty-foot ocean waves were a majestic demonstration of the unseen force that powers everything. Catching a glimpse of something that elemental, that beautiful, and that powerful created one inevitable result: the desire to see it again
>
> (2011: 19).

For surf-riders, their irruptive moment in surfing spaces affectively transforms them, exerting upon them a pull to experience that co-constitution repeatedly.

The energies transferred to them through surf-riding stimulates further engagement, as the following surf-riders state, engaging with surfing space becomes:

> a compulsion, an addiction, what I live for and can't live without. It keeps me in a constant state of restlessness, my mind and body always searching for the next wave
>
> (Britton, 2015: 118).

> [surfing is] a seriously addictive passion. Initially I had assumed that surfing once a week or fortnight would be more than sufficient. Yet the more I surfed, the more I wanted; the hungrier I became
>
> (Capp, 2004: 82).

And as American writer and surfer Steven Kotler states:

> Once you've had a taste of [surfing spaces, it is] something different, something of out there, [and] then it's hard to give it up. [It] gets its hooks in you. Afterwards nothing else can make you feel the same (2006: 153).

And as Leza puts it:

> Surfers ... need their fix, if the government knew how addictive surfing is they would have outlawed it by now (Leza, 2012).

These quotations demonstrate that for many surf-riders, surfing spaces realise the "active transformation of the human being" (after Feltham and Clemens, 2003: 7), regenerating them as littoral-humans who have an embodied connection to surfing spaces. This compulsion (see also Anderson and Stoodley, 2018) in turn prompts individuals to re-orient their lives to some degree or other, to enable this continued engagement with surf-riding. This "fidelity" comes in many forms; from finding time in a busy life to surf-ride in the evenings or weekends, or fulfilling a promise made to a future self to indulge in an exotic surf-trip, as Tom Anderson narrates with respect to his vicarious experience of J-Bay that was noted above:

> ... what natural-foot[3] serious about surf travel wouldn't dream of making a pilgrimage to that little stretch of coastline just beyond Humansdorp? This trip for me had begun over ten years ago [in that irruptive moment in the Porthcawl pub], with a decision. A promise. An *oath*. "Thou shalt surf J-Bay once thou hast grown old enough to travel there and once thou surfest good enough to make it count"
>
> (Anderson, 2007: 9).

The fidelity shown to the event of surfing spaces can therefore come in many forms, yet as we will see in Part II, a surfing spaces script has emerged which encourages a particular form of dedication and commitment. Duane's excerpt above indicates this idealised commitment (when it is assumed that any "true" surf-rider would find it impossible to combine their fidelity to surfing spaces with becoming President of the United States of America). The ideal dedication is thus suggested to be an obsessive form of fidelity which involves the re-orientation of one's whole life so that surf-riding can occur whenever and wherever conditions allow. The following excerpt from *The Surfers' Path* exemplifies how surf media often impose and consolidate certain interpretations of fidelity, as they respond to the question "how do you know you're a real surf-rider" by prescribing:

> Well for a start, [surf-riders] have probably made some major, life changing decisions around the need to live near the ocean and to somehow create enough freedom to surf every day good waves are happening. [They] probably can't look at any piece of water, north or south coast, lake, river or stream, without searching every detail of the water surface, the bottom contour, wind and swell for the sign of rideable waves. Even if they're six inches not six feet, there's still something irresistible about them. [Their] partner probably has a problem with [their] obsession and passion for the sport: "If only you were into me as much as your damn board!" [Surf-riders] can't contemplate never being able to surf again. The prospect of the next great day and [their] next great wave is one of the things in life that keeps [them] going forward. Moving away or just not being physically capable of surfing are prospects that just don't bear thinking about. [They]'re always restless, dream[ing] of the next session, [their] optimism tells [them] that the next swell will be the best ever, and [they]'ll be on top of [their] game when it comes. On the days it's working but [they] miss it, [they] get eaten away inside with frustration. [They]'ve probably bunked off school or work many times because it really was too good to miss. Driving to good surf, hurriedly changing and running headlong to the water, fills [them] with stomach tightening anticipation, the same as it ever did
>
> (Richards, 2002: 114).

This overwhelming sense of fidelity is well summed up by surfer-writer Allan Weisbecker when he explains how, for many surfers, the eventful nature of surfing spaces,

> deeply influences every aspect of our existence, from where we live, to the cars we drive, to who our friends are, to the temperament of a prospective mate, to our career goals, if any manage to surface through the turbulence of our single-minded passion for the water. But more than anything, surfing forges our perception of ourselves and of our relationship to the world around us
>
> (2001: 11).

Surfing spaces can therefore be seen to have "eventful" effects on many who experience them. These events transform terrestrial individuals into geo-humans, defined and co-constituted by littoral locations. These (now) littoral-humans feel compelled to experience again and again the relational sensibilities produced through being part of surfing spaces. They demonstrate fidelity to their original irruptive moment by re-orienting their lives (to various degrees) to make this possible and therefore be defined, for fleeting moments, as a surf-riding cyborg.

Conclusion

Identifying surfing spaces as eventful is significant in the context of this book as it not only faithfully re-presents the assembling of surfing spaces that we have examined in Part I, but also foreshadows the cultural practices that have scripted surfing spaces which we will examine in Part II. It serves to emphasise how surfing spaces are extraordinary phenomena, and explains to some extent why the constituent act of surf-riding often appears as a cultural geography and practice to which many aspire: surfing spaces are irruptive, addictive, and fun – who would *not* want these experiences in their lives? However, identifying surfing spaces as eventful also reminds us that such aspirant cultural geographies and practices are always subject to (b)ordering, to the framing, commodification, and even exploitation of their constitution and definition by particular cultural groups, including cultures and tribes of surf-riders. In Part II we examine these (b)ordering processes directly, contributing to Brubaker's call to not only "give serious theoretical attention to [the composition of] contingent events [but also] their transformative consequences" (1996: 21). Part II documents how, in the twentieth century, one particular version of the surfing space assemblage came to prominence, successfully imposing its own ontologies, fidelities, and technologies – in short its own "scripts" and "codes" – to determine how surfing spaces are understood globally. Yet as is the case with all (b)ordering processes, attempts to define and normalise do not occur without challenge. Part II thus not only demonstrates the rise to dominance of particular surfing space scripts, but also how they are being actively resisted, subverted, and transgressed to influence and define surfing spaces in the contemporary period. Part II begins by exploring the heterogeneous histories of surfing spaces.

Notes

1 This point is reflected in the oft-quoted examples that an "event" can be the moment of "falling in love" (see Bensaid, 2004: 98). As Bensaid puts it, "the genuine event … is of the order of an encounter that is amorous (love at first sight), political (revolution), or scientific (eureka)" (2004: 98)
2 It is possible to watch Curren's ride for yourself here: https://www.facebook.com/StabSurfMagazine/videos/10156986581725408/
3 A "natural foot" is a right-footed surfer. A "goofy foot" is a left-footed surfer. Some wave breaks, like that at J-Bay, favour the natural foot stance.

Part II

Part II

9 The Heterogeneous Histories of Surfing Spaces and their Cultural Colonisation

Introduction

As we have seen in Part I, surfing spaces are extraordinary phenomenon. Although dependent in part on physical processes, the geographies of surfing spaces are also assembled with and through human cultures, bodies, and surf-riding technologies. The cultural geographies and practices of surfing spaces are therefore not wholly natural or neutral in their composition, like all spaces their "gatherings, weaving, and assembling are… subject to, and productive of, the influence of power" (after Cresswell, 2019: 197).

Surfing spaces are thus composed in part through a process of cultural ordering and geographical bordering, where one cultural group's ideas of "good" and "bad" or "right" and "wrong" are transformed from being merely one way of being into the only way to be. Through this process, cultural ideas are realised in physical form as they take and make the geographies of surf-riding. In Part II we examine these (b)ordering processes directly and document how, in the twentieth century, one particular version of the surfing space assemblage came to prominence and successfully determined how surfing spaces are understood globally. It begins by providing important insight into the (under-acknowledged) historical heterogeneity of surfing spaces, and progresses to account for the rise to power of one iteration of this assemblage. It does so through re-counting the more well-known marginalisation and appropriation of endogenous Hawai'ian surfing through state- then entrepreneurial-colonization. These insights provide the basis for what this book argues is the "scripting" of surfing spaces in the 1950s and 1960s, and which is explored in detail in Chapter Ten.

Pre-colonial surf-riding

When researching the history of surfing spaces, scoping decisions are always framed by the influence of power. Theoretical predispositions, philosophical fashions, and even vested interests serve to mobilise thinking and frame interpretation, and, as a consequence, any account of surfing history may, "conceal[…] as much as it reveals" (Mansfield, 2009: 12). Scoping decisions about the inclusion or exclusion of sources (be they primary or secondary),

DOI: 10.4324/9781315725673-9

alongside the degree of transparency of and accountability for those decisions, are crucial insights so the veracity of any account can be judged by the reader. In relation to the primary stakeholders included, it is possible to reflect on the authority they may have, in what subjects, and whose interests they are representing; in relation to secondary sources, we may reflect on what types of evidence may be valorised as "good knowledge" (perhaps written sources documented by "experts" for example), and also what alternative sources may be marginalised as a consequence (perhaps oral accounts by those deemed "non-experts"). As a consequence, it is possible for historical and contemporary colonising processes to silence some insights into surfing history if they do not coincide with prevailing predispositions towards authority and interest[1]. It is in this context that the histories of surfing spaces are composed.

> What [passes for the *definitive* history of surf-riding] today… [is actually] surfing as profiled by the *popular* history of the sport
> (Westwick and Neushel, 2013: 2, my emphasis).

> Popular histories "stand as an unavoidable source base in their purposeful recording of modern surfing history… [yet] are riddled with silences, omissions, and distortions"
> (Hough-Snee and Eastman, 2017: 9/10).

As Westwick and Neushel, then Hough-Snee and Eastman suggest, we must be careful when relying on popular accounts of the history of surfing spaces. Although these accounts often emanate from individuals who have a high degree of first-hand experience in the surfing spaces of which they are a part (and as Chapter Ten outlines in more detail, have often played in a key role in scripting surfing spaces in the twentieth century), these histories often rely on narrow definitions of what surf-riding is and should be; and due to the synergistic relations between the storying, control, and commodification of surfing spaces, these historians may not always actively seek to offer or support disputing accounts[2]. As a consequence, as Hough-Snee and Eastman note, although these modern accounts may be "unavoidable" due both to their omnipotence and insight, they are nevertheless inevitably partial.

This partiality is also due to popular accounts relying on a small cadre of practitioners and writers. As Zach Weisberg, former Editor of *Surfer* and founder and Editor in Chief at *The Inertia*, notes, in the popular surf media, it is common to "only hear from individuals from Orange County or Gold Coast, or kind of these surf industry meccas…" (Research Interview), whilst Hough-Snee and Eastman concur, suggesting that the history emanating from surf media remains,

> an exclusive affair limited to small circles in specific geographic areas (Southern California and Australia's east coast, for example). Outsiders to

these circuits, and particularly women, are noticeably absent from the ranks of mainstream surf media

(2017: 10).

Popular surf histories also remain inevitably partial due to what the surf media themselves describe as a "romanticism" at the heart of the spaces they produce. As Luke Gartside, Editor of UK-based surf magazine *Wavelength* suggests,

> Surfers'... fascination with heroism, peril and the transient and ferocious nature of our playground combine to create fertile ground for all sorts of grand stories [to such an extent that] romanticism [plays a] distorting role in our sport's official history
>
> (2020: 6).

Those who write the popular (or as Gartside terms it, the "official") history of surfing spaces thus acknowledge that, to some extent, these versions are distortions, their preference for a romantic narrative trumping a commitment to verifiable fact, or as surf journalist Brad Sterling puts it, these authors "never let the thirst for truth get in the way of a bloody good yarn" (2000: 40). Regardless of whether this partiality should be benevolently framed as romanticism or otherwise labelled as self-interested culturalism, it is clear that instead of prioritising accuracy or rigour, when framing these histories this privileged cadre,

> prefer to shape, design, and choose their collective past. [...] As one surf journalist wrote in 2005, "Ours has always been a culture of storytellers, not historians"
>
> (Warshaw, 2010: 22)[3]

Thus seeking to document what we might call "pre-colonial surf riding" is a daunting exercise. When others have sought to do so, they have done so with the idea that any "pre-colonial" surfing spaces are a diverting by nevertheless disposable "preface" to the main, most important, story (as Warshaw states in relation to Pomar's "obscure" history, any wave-riding in ancient Peru should really be sidelined as "a self-contained prelude to surf history, not the starting point" (Warshaw, 2010: 22). Thus we must take care not to perpetuate the assumption that the popularised surf history is the only (valuable) one, and that in re-telling some or all of this story, we acknowledge some degree of complicity in sustaining its power-laden definitions, values, omissions, and myth-making. In this spirit, this chapter continues by acknowledging the range of indigenous surf-riding practices that can be identified in academic and popular surf-riding literatures, whilst also accepting that these literatures will not be exhaustive due to their (and this book's) reliance on written histories, which in turn may be limited by their own omissions and silences too[4].

With all this in mind, recounting "pre-colonial" surfing spaces serves a number of purposes. To draw attention, if its required, to the facts that surfing spaces predate (and may run in parallel to) contemporary modern versions; that pre-colonial versions were (and may still be) diverse in their geographies and cultures; and finally, that contemporary versions of history are de facto colonial versions, whether we find that label appealing or otherwise, and thus it is necessary to critically reflect on their (b)orders and consequences.

Everything you know is wrong (?)

Hawaii invented surfing

(Warshaw, 2010: 71).

Surfing expanded from the U.S. to the rest of the world

(Esparza, 2016: 199).

[Popular surf historians] generally believe the first account of surfing was written in Hawai'i in 1778. They are only 140 years too late, and some ten thousand miles off the mark

(Dawson, 2017: 139).

If we step outside the popular history of surfing spaces, we can identify that often independent, indigenous surf-riding practices occurred in a wide range of locations. Esparza (2016) defines these surfing spaces as those with an "endogenous genesis", in other words, cultural geographies which are free from "foreign influence" in their origin. Whilst Esparza suggests that such endogenous cultures are limited to Polynesia, others point out that "aboriginals all over the planet were being propelled by waves long before [explorer] intrusion" (Stecyk, 2002: 32). As Dawson argues therefore, although conventional wisdom suggests that the key location for the endogenous genesis of surf-riding was Hawai'i, he nevertheless documents that surf riding,

was independently invented throughout Atlantic Africa[5] and Oceania [and the] first account [that could be secured] was written during the 1640s in what is now Ghana

(2017: 139).

Dawson suggests that in these locations, surf-riding was undertaken in "a prone, kneeling, sitting, or standing position, and in one-person canoes" (ibid). Hough-Snee and Eastman identify Hemmersam's and later Alexander's 1834 accounts of (west) African surf-riding, where:

From the beach ... might be seen boys swimming into the sea, with light boards under their stomachs. They waited for a surf; and came rolling like a cloud on top of it

(2017: 140).

There is evidence to suggest that, in line with Pomar's account (above), surf-riding also endogenously existed in South America. As Stecyk states,

> Archaeological evidence dating back five thousand years at Chan Chan in Peru depicts men [sic] riding waves. These prehistoric glyphs and related ceramic sculptures show surfers riding the caballito[6] (little sea horse), a short craft made of bound reeds, with an upturned end, which is still in use today as a fishing and surfing craft

(2002: 34).

The caballito rafts were not dissimilar to canoes (perhaps 9ft in length), were made from the *totora* plant, and were surfed using a paddle.

Indigenous forms of surf-riding also existed in many islands in Polynesia (including but not limited to Hawai'i). Stecyk suggests that surf-riding in Hawai'i dates from at least AD300 (and perhaps even earlier) when the Maoli arrived from the Marquesas Islands and brought the practice with them (Stecyk, 2002: 39). Throughout Polynesia, surf-riding was practiced by all members within society, with a variety of styles (prone and sat) and technology (bodysurfing, canoes, and boards) employed; but it was only in Tahiti and Hawai'i where one choice among many was to ride "full-length boards while standing" (Warshaw, 2010: 23). As Warshaw suggests,

> Hawaiians developed three basic types of surfboard: the paipo, the olo, and the alaia. The round-nosed paipo was the smallest of the three and was used mostly by children in nearshore surf. A child's paipo might be 3 feet tall, 16 inches wide, and a half-inch thick. An adult version could be as long as 6 feet. [Some] could stand on a paipo board, and a kneeling position was used, too, but mostly it was ridden prone

(ibid: 25).

In Polynesia, this diversity of surf-riding was undertaken by all inhabitants of society, as Finnegan writes,

> ... Men and women, young and old, royalty and commoners surfed. When the waves were good, "all thought of work is at an end, only that of sport is left," wrote Kepelino Keauolalani, a nineteenth-century Hawaiian scholar. "All day there is nothing but surfing. Many go out as early as four in the morning." The old Hawaiians had it bad, in other words – surf fever

(2015: 31).

From such accounts, modern day scholars surmise that Hawai'ian and other pre-colonial surfing assemblages were far from being conflictual spaces, as Hill suggests, prior to its modern incarnation, "surfing's norm has been inclusion" (Hill, 2020: 32). This hypothesis may be likely, and is a sound basis for argument based on histories of Polynesia in particular (in Hawai'i at least there seems to be broad consensus that surf-riding was a socially inclusive and polyvocal activity). What these accounts do demonstrate, however, is that surfing spaces existed in many locations, in many styles, using many technologies, before what has been termed "modern" surfing (Warshaw, 2010) – what we may term colonial surf-riding – was "invented" (after Finnegan, 2015: 31). Pre-colonial and endogenous surf-riding existed (and may still exist in some places), and thus surfing's history is indisputably heterogeneous in nature. In this context, the dominant surfing spaces of the twentieth and twenty-first century may now be understood as resolutely colonial in nature, and as we will see in the next section, originated from the violent appropriation and subsequent commodification of indigenous surf-riding practices in Hawai'i by religious-, nationalist- and capitalist-colonisation.

Indigenous surf-riding meets empire, colonisation, and modernity

So far in this chapter we have seen that endogenous surf-riding cultures existed in many places around the globe. This chapter continues by broadly charting the emergence of one particular version of the surfing space assemblage from its origins in the colonial dismantling of indigenous Hawai'ian surfing spaces, to the subsequent re-imagination of them through the lens of colonising powers. The chapter suggests that the initial phase of this colonised assemblage can be understood as not simply a form of cultural imperialism, but also cultural entrepreneurship, injecting specific values (including individualistic, capitalist, and to some extent racist and patriarchal ideals) into an imaginary of Americanised-Hawai'i. We will see that there were three key phases in the development of this assemblage up until the mid-point of the twentieth century. Firstly, the initial conquest and colonisation of Hawaii's cultural geographies; secondly the subsequent entrepreneurial recuperation of indigenous surfing spaces in order to appeal to (white) tourist markets; and thirdly, the creative colonisation of this recuperated assemblage through adaption with Californian culture. These phases set the foundation for the "scripting" of surfing spaces which occurred in the second half of the twentieth century. This chapter progresses to detail the key phases noted above, before Chapter Ten critically explores the rise to prominence of the surfing spaces script.

Colonisation through conquest

As argued elsewhere (see Anderson, 2021), technological and military advancement in the eighteenth century heralded an unprecedented expansion of the geographical, cultural, and political power of a small number of nation states. This was a process of empire building through external colonisation. Following Fenster and Yiftachel (1997), external colonisation involved one state invading another, sometimes peacefully, but often violently, subjugating and taking control of foreign tribes, their land, culture, and resources. Through this process states, "imposed political and military control, battled or bargained with local ruling elites, [and] confronted or diverted local opposition" (Godlewska and Smith, 1994: 56). This (b)ordering process was imposed upon the cultural geographies, including the surfing spaces, of Hawai'i.

> "[British explorer] Captain Cook's landing at Waimea on Kauai in January 1778 brought the alien forces of the outside world into collision with Hawai'i's isolated Polynesian society. The intrusion of western European and American colonial forces rapidly destroyed or greatly diminished traditional Hawai'ian activities such as surfing"
>
> (Stecyk, 2002: 38).

Cook's arrival in Hawai'i marked the beginning of the colonisation and enclosure of its culture, geography, and economy by western powers. As The US Department of State's official archives outline,

> For most of the 1800s, leaders in Washington were concerned that Hawai'i might become part of a European nation's empire. ...In 1842, Secretary of State Daniel Webster sent a letter to Hawai'ian agents in Washington affirming U.S. interests in Hawai'i and opposing annexation by any other nation. ...In 1849, the United States and Hawai'i concluded a treaty of friendship that served as the basis of official relations between the parties. ...When Queen Liliuokalani moved to establish a stronger monarchy, Americans ...deposed her in 1893. ...Spurred by the nationalism aroused by the Spanish-American War, the United States annexed Hawaii in 1898 [and] Hawai'i was made a territory in 1900
>
> (US Department of State, 2001–9: no page).

During this invasive transition from settler arrival to formal annexation, white settlers not only brought disease to Hawai'i (decreasing the local populations from 800,000 to 40,000, see Stannard, 1989), but also imposed a capitalist economy which stripped locals of their leisured time. The indigenous cultural geographies were thus placed under pressure from western interests, as Finney outlines:

> During the century of tears that ended with the overthrow of the Hawai'ian monarchy by foreign (mostly American) residents backed by the U.S. Marines, the Hawai'ian language and many cultural practices were driven underground, and some were lost altogether
>
> (2002: 91).

Yet to some extent surfing spaces remained one aspect of Hawai'ian cultural geographies in which indigenous people retained their identity and expertise, as Hawai'ian chronicler Isaiah Helekunihi Walker points out:

In the ocean, Native surfers secured a position on top of a social hierarchy. Because Hawai'ian surfers contended for this autonomous cultural space they had the freedom to defy colonial prescriptions for how Hawai'ian men should behave. As they transgressed *haole* expectations and categories in the waves, Hawai'ian surfers simultaneously defined themselves as active and resistant Natives in a colonial history that regularly wrote them as otherwise

(in Warren and Gibson, 2014: 33).

Yet, as with other aspects of Hawai'ian culture, due to the opportunities for resistance and transgression that surfing spaces afforded, these locations were also (b)ordered by incoming elites and surf-riding was formally banned (see Peralta, 2004: 6.25). Although tight control of all aspects of Hawai'ian cultural geographies led to the decline of surfing (as quotations suggest above), its eradication was not absolute. As Helekunihi Walker has shown, in the context of ongoing resistance from the Hawai'ian people,

surfboard[s] were being made outside the Western cash economy and beyond the tourist gaze, as Hawaiian surfer and board maker Tom "Pohaku" Stone explained:

"Hawaiians never stopped surfing or making surfboards like the [surf] magazines and books claim. The thing is our society did not develop around selling things for green bits of paper, and we didn't make surfboards in a shop. But we did craft them. People coming here may not have seen this going on and so Westerners start thinking they have reinvented surfing and surfboard making in Hawai'i. But we were surfing and making boards the whole time"

(in Warren and Gibson, 2014: 36).

Colonisation through cultural entrepreneurship

During the nation-building project which ended in Hawai'ian annexation in 1898 (and full assimilation to US statehood in 1959), the cultural resources of Hawai'i became subject to the entrepreneurial gaze of incoming Americans. Although many with a political or religious zeal viewed much of Hawai'ian culture as threatening to moral (b)orders, others saw in them an opportunity for profit. If these cultures could be translated in line with the tastes and ideals of US society, it would be possible to open up Hawai'i's cultural geographies to new tourist markets. In relation to surfing spaces, this was led by the entrepreneur Alexander Hume Ford. As Peralta summarises (again, note the popular "distortion" of events when compared to Helekunihi Walker above):

The extinct Polynesian pastime [of surfing] was then reintroduced in the early 20th century by Alexander Hume Ford, globetrotting promoter who set about reviving Island tourism by romanticising surfing at Waikiki

(2004: 6.42).

Through Ford,

> surfing was used to sell Hawai'ian tourism – and white settlement – in popular magazines and promotional literature, in the process strengthening the grip of the haole class over the native population
>
> (Laderman, 2014: 3).

In one sense this was a first indication of how cultural entrepreneurship could re-organise the "cultural, financial, social and human capital" of surfing spaces "to generate revenue" for particular cultural groups (after Anheier and Isar, 2008: 1). In this case, the indigenous activity of surf-riding was recuperated, a process of appropriating the "antagonistic expressions [of a culture] and render [ing] them harmless through transformations and integration into some form of commodity" (Aufheben, 1996: 34). No longer were the surfing spaces of Hawai'i characterised by plurality and conviviality between crafts and riders, they were now domesticated in line with the values of American settlers. In this way not only did cultural entrepreneurship successfully commodify what remained of endogenous culture, but as Laderman states above, consolidated through recuperation the control over indigeneity that settlers had established. Thus Hawai'ian surfing spaces were translated into privileged destinations for rich, white America. It is also significant to note that this latter objective was not simply entrepreneurial or nation-building in its nature; Hume's idea(l)s were informed by his support for the segregation of America, and he sought to impose racist (b)orders on the surfing and other spaces of Hawai'i, as Dawson states:

> During the early twentieth century, Alexander Hume Ford and Jack London sought to marginalize Hawaiians while using surfing to promote tourism. A South Carolinian transplant to O'ahu and product of the Jim Crow South, Ford "embraced the racist suppositions of the post-Civil War era"
>
> (2017: 147).

Indeed, as Laderman reports,

> Ford penned a sequence of enthusiastic articles for *Van Norden Magazine* intended to entice white migration. "Hawaii is to-day the land of opportunity for the quick, active, courageous white man, and every one from President Taft down wishes to see it conquered for and by Anglo Saxon Americans," he proclaimed. In a piece entitled "Hawaii Calls for the Small Farmer," Ford insisted that the "richest land in all the world... must be Americanized"
>
> (Laderman, 2014: 28).

As part of the process of cultural colonisation by entrepreneurship, expert surf-riders (including "legendary lifeguard George Freeth and the five-time Olympic medallist Duke Kahanamoku" (Laderman, 2014: 3) were encouraged to promote

Hume's version of Hawai'ian surfing spaces to potential markets across the globe. In this way Freeth and Kahanamoku became "surfing's Johnny Appleseed[s], introducing [their] favourite sport to far-flung places like California, New York and Australia" (Peralta, 2004: 7.14[7]). The promotion of this Hawai'ian surfing assemblage was key in attracting individuals to Hawaii to not only holiday and invest in this newly Americanising space, but also to further develop the version of surf-riding that Hume had set in motion.

Colonisation through cultural creativity

> One of the fans enthralled by the Duke was a young Wisconsin swimming champion named Tom Blake. Relocating to Hawai'i, Blake would go on to become one of the 20th-century's most influential surfers through his innovative surfboard design, but most importantly, through his advocacy of surfing as a way of life
> (Peralta, 2004: 7.22).

The colonisation of endogenous Hawai'ian surfing spaces was thus initiated through invasion then entrepreneurship. But this recuperated and commodified version of surfing spaces was in turn developed through those who visited Hawai'i as a consequence of its promotion. Occurring in the relatively peaceful but economically depressed eras of the 1920s and 1930s, this development occurred not through continued invasion or commodification, but through cultural creativity, and a key exponent of this process was Tom Blake.

Living in both California and Hawai'i, Blake used innovation and creativity to develop surfing spaces. He experimented with surf photography including waterproof housing and pioneered new ideas for craft technology to enhance manoeuvrability and agility on the wave. Blake devoted all his time to surf-riding and surf-related innovation, and documented his passions and philosophies in this regard (1935); with his laid-back "sun-bleached, rumple[d] personal style [he] became the prototypical beachcomber [, a] look... still in effect today" (Lawler, 2011: 70). As Blake himself summed up:

> I found the water good... better than the land I was cut off from. Water supports a rebel, if he has the will and ability to swim
> (Warshaw, 2010: 65).

As Warshaw suggests (and we will see in Chapter Ten), surf-riders and subsequent entrepreneurs "would pull as much utility from this Blake-defined notion [of surfing spaces] as they would from the hollow board, or the fin, or the surf photo" (ibid.: 65).

Thus through innovation and creativity surfing space assemblages changed. The endogenous surfing spaces of Hawai'i that were marginalised through invasion and had been further sidelined through entrepreneurial commodification, had now mutated through the impetus of ideas and practices from California. These locations and influences parented a new cultural and geographical chimera (or "bastard child" as Alf Anderson describes it (2011: 128)) of surfing spaces. As Warshaw states above, it was this Blakean-influenced assemblage that the post-war generation of surfers in California would "pull as much utility" from as possible, uncritically adopting a cultural amnesia over the erasure and silencing of endogenous Hawai'ian surfing spaces and repackaging them into their own version of surfing spaces. As we will see in Chapter Ten, from this basis, littoral locations would become scripted as a place of freedom and escape for the adventurous individual; surfing spaces would become the new American frontier.

Notes

1 For example, definitional disposition can include or exclude insight; if surf riding is considered narrowly (i.e. just stand up surf boarding) then this can change the course of published history (see Esparza, 2016). If narratives are selected from similarly narrow definitions, then his-storying surfing spaces can become accounts of privilege rather than inclusion, with the silencing of non-literate, non-white, non-male, and otherwise non-privileged identity positions. As Dawson argues, in this way, it is likely that "History is typically studied through the Western lens [as] Westerners ... physically and intellectually colonize [other] ...cultural understandings of amphibious spaces" (2017: 149).

2 This point is well made by Hough-Snee and Eastman when they state,
"today's most prolific popular surf historians, Matt Warshaw and Drew Kampion, cut their teeth as writers and editors at major surf magazines. Kampion is a prime example of this insider positionality in the surf industry, where writers wore multiple hats as practitioners, competitors, business owners, advertising agents, popular historians, or all of the above. An editor of *Surfer* from 1968 to 1972 and *Surfing* from 1973 to 1982, Kampion then headed the advertising department for apparel brand O'Neill, before penning several foundational [surfing histories]" (2017: 10).

3 This point is perhaps best illustrated by Warshaw's own account of the meeting between Peruvian surf-rider Felipe Pomar with *Surfer* Magazine in California in 1987. Pomar suggested he had evidence for surf-riding originating in Ancient Peru. As Warshaw states, Pomar argued that,
"some five thousand years ago – the fluttery thrill of riding a wave became its own reward [for Peruvian fishing communities]. This easily repeatable and wholly non-productive act was then removed from the daily work routine and pursued for its own sake. A form of surfing began. The *original* form" (2010: 21).
As Warshaw continues:
"A simple lack of interest in obscure historical debate doesn't fully explain the booming silence that greeted Pomar's 1988 *Surfer* article. It was defensiveness, too. Surfers love the idea that their chosen activity was born in translucent blue water, next to palm-fringed beaches, and practiced by royalty on beautiful wooden surfboards. It's the "Sports of Kings", and even if the phrase was created by some early version of the Waikiki tourist board [as we will see later in this chapter], most surfers nonetheless wear the designation with quiet pride. Good luck trying to sell the idea that reed-boat-straddling Peruvians trolling for anchovy off the grim brown coast of Peru were the *real* first wave riders. ...when it comes down to Hawaii or Peru, the tropics or the desert, the Sport of Kings or the Sport of Fishermen -well, that's hardly a choice at all" (2010: 22).

4 As C R Stecyk suggests, surf-riding was "an act of cultural and religious significance for [many] extinct societies [who] left us no written records" (2002: 32), and thus as a result, ran the risk of their knowledges being lost to historians that valued other modes of knowledge and expertise.

5 Including, suggests Dawson, in "Senegal, the Ivory Coast, Liberia, Ghana, Cameroon, and West-central Africa" (2017: 139). Finney identifies similar accounts, concluding that despite "West African surfing [being similar in nature to] paipo surfing [in Hawai'i], I found no evidence that it had spread to West Africa from Hawai'i. Moreover, the West Africans used crossbars to grip the boards, a feature not found in the Pacific. These factors led me to conclude that West African surfing probably developed on its own" (Finney, 2002: 85).
6 Caballito is the word used in Spanish to refer to this craft, whilst "mochica tup" is used in the indigenous language (see Hough-Snee and Eastman, 2017: 2).
7 It is of course questionable whether this promotion "introduced" surf-riding to these locations (see for example Sterling, 2020), but in line with the argument of this book, it is clear that it did "introduce" this *new* version of the surfing space assemblage to these different locations.

10 The Imagineering of Surfing Spaces

Introduction

As we have seen in Chapter Nine, surfing spaces have heterogeneous histories with a range of endogenous surf-riding cultures existing around the globe. However, the popular histories of surfing spaces often focus attention on the development of one particular surfing space assemblage based at the littoral location of Hawai'i. As we have seen, this endogenous surfing space was colonised through cultural invasion, then recuperated through capitalism, and developed through cultural creativity (Chapter Nine). From the mid-point of the twentieth century however, ideas and practices associated with a *new* version of this surfing space rose to prominence and, within a decade, had come to be the hegemonic way in which surfing spaces were understood globally and would remain so until the present day.

The next two chapters will argue that this new surfing space occurred due to the selective re-packaging, reproduction, and representation – or "imagineering" (after Nijs and Peters, 2002) – of key aspects of Californian-Hawai'ian surf-riding up to that point. Drawing on the vocabulary of Melucci (2006), this chapter argues that the process of imagineering successfully transformed the energy of wave-riding into a "script". A script is a meta-narrative or central framing which is often symbolic in nature and created to "mobilise", "validate", and "enlarge" (after Gamson and Wolfsfeld, 1993: 166) interest and participation in a particular cultural geography. Following Melucci, this script is constituted in practice by range of cultural "codes". Codes constructively align with the idea(l)s of the script and define the specific (b)orders of its cultural geographies. The imposition, adoption, and subsequent perpetuation of codes complete the imagineering process as prescriptive (b)orders of access, entitlement, and marginalisation are realised in practice.

In Chapter Eleven we outline the consequences of the surfing spaces script. We explore its success firstly in America and then overseas, and how the imagineering process not only invented the role of "the surf-lifestyle entrepreneur" (a group of people who could make a living from moulding and shaping the activities associated with the surfing spaces script), but also the "surf-rider identity" (a set of surf-related activities around which a life could be defined).

DOI: 10.4324/9781315725673-10

The remaining chapters in Part II (Chapters Twelve to Sixteen) examine the "codes" that have produced the dominant version of surfing spaces in practice. In this chapter however, we detail the core elements of the surfing spaces script as it was imagineered in the 1950s and 60s. We will see how a particular script was imagineered by a small group of young Californians who did not wish to sacrifice their surf-riding in order to gain conventional waged employment. To solve this problem, they attempted to make a living from selling their experiences of surf-riding, and to this end, they imagineered a script which not only introduced representations of the surf-riding experience to new audiences, but also directly connected them to dominant western idea(l)s of individualism, freedom, and the open frontier. In this script, the surf-rider was framed as a heroic protagonist who, through his brave and dramatic encounters with surfing space[1], was able to escape the confines of mainstream culture. This script suggested that such escape could not simply be found in the instant of surf-riding, but also in the broader practices associated with surfing spaces. In this framing therefore, individuals became scripted as idiosyncratic adventurers always living a personal "quest" for "stoke". The script had the effect of elevating all aspects of this surfing space assemblage from the realities of the contemporary, and appealed strongly to the zeitgeist of both the "rebellious" youth culture of the 1950s, and later the counter-cultural movements of the 1960s.

As this chapter examines the imagineering and scripting of surfing spaces, it also introduces the media architecture which disseminated its meaning and values. As Horton identifies, "scripts" are enacted through a broader "architecture" that serves to prescribe and perpetuate ideas in action (2003). Developing from stories and oral narratives amongst individuals in littoral locations, and expanding to include modern media outlets (including the popular consumption of magazines, films, music, and merchandise – which would themselves rapidly proliferate across the globe in the latter half of the twentieth century), an array of words, images, and other commodities were harnessed to communicate the surfing spaces script initially throughout Western America, and then to key surfing markets around the world. We begin this exploration of the surfing spaces script, it's values and architecture, through introducing the process that created it: the practice of "imagineering".

Imagineering a new way to live

> How do new ways of living come into the world? How do we find new understandings of the right or good way to live?
>
> (Jordan, 2002: 7).

This chapter argues that in the 1950s, a new form of "surfing" came into the world through a process of "imagineering". Imagineering is a compound word which merges "imagination" and "engineering", and reflecting the broader relational approach of this book, "consists of two existing nouns, which are then adapted in such a way that a new word and meaning emerge" (Kuiper and Smith, 2014: 8). Resonating with the relational capacities of all surfing spaces, imagineering (not simply as a word but a practice) brings once separated ideas and actions together, and through so doing, becomes a generative verb. To imagineer, therefore, is to do more than simply "construct"; "construct" involves using intellect to impose order, whereas to imagineer goes further. To imagineer employs fiction, imagination, folklore, and oral narratives, along with culturally significant ideas and values, to impose order on and give meaning to selected elements of reality. An imagineered world seeks to appeal to its constituency (which are predominantly and importantly positioned as "consumers" by those imagineering it, see also Chapter Eleven) creating not simply functional products or practices, but "meaningful experiences" (after Kuiper and Smit, 2014: 1) which align "on a rational *and* emotional level" with the target culture (Nijd and Peters, 2002, cited in Kuiper and Smith, 2014: 8, my emphasis). Imagineering also goes beyond the "simple" construction of abstract ideas due to its definitive need to be "engineered" or put into practice; as Routledge et al state: imagineers, "work to effect what Callon ... terms the 'moments' of translation" (2007: 2578), turning not only ideas into action, potential consumers into a functioning market, but also disparate individuals into a culture. In sum, "'imagineers'... 'ground' the concept or imaginary ... (what it is, how it works, [and] what it is attempting to achieve)" in context (Routledge, et al, 2007: 2578).

The process of imagineering acknowledges that, despite any encounter or experience being personally "eventful" in nature (as we have seen with respect to surf-riding in Chapter Eight), the imposition of meaning can re-position these encounters within a broader set of symbolic values, as Gamson and Wolfsfeld put it,

> events do not speak for themselves but must be woven into some larger story line or frame; [if this occurs] they take on their meaning from the frame in which they are embedded
>
> (1993: 117).

The process of imagineering provides this "larger story line or frame" to eventful encounters, and this chapter will demonstrate how, with respect to surfing spaces, the appeal of littoral locations was woven into a "script" shaped by the cultural commoditisation of surfing generated by Hume, the innovations produced by Blake, as well as the interests and values of a small group of surf-riding Californians in the 1950s. This script provided a "compelling" framework which imposed meaning and value on the surf-riding experience (after Bennett, 1975: 65), one which remains powerful to this day.

Imagineering the dream: A new surfing spaces script

> The dream was invented by one generation of young, mostly white, Californian men...
>
> (Comer, 2010: 36).

> The original guys, from southern California, had a very narrow view of what surfing should be. This [narrow view] has carried over into [the present day]
> (Elizabeth Pepin Silva, surf photographer and film-maker, Research Interview).

The imagineering of a surfing spaces script in the 1950s was not a natural or inevitable phenomenon. It was a working solution to a practical problem faced by a loose coalition of relatively young Californian surf-riders. A few individuals, including but not limited to, Severson, Braun, Edwards, Alter, Stoner, Brewer, Ball, Grannis, Noll, Copeland, Brown, and later Kampion, understood the relational sensibilities of surf-riding. Through their own practices they had experienced the fun, thrill, and risk of surfing, the sense of disconnection it offered from the pressures of life-on-land, and the re-connection it promised to broader relational sensibilities of self-affirmation, convergence with nature, and spirituality (see Part I). These individuals had experienced first-hand the sense of exploration when encountering waves in general, and the promise of adventure generated through the discovery of new waves (to them at least) just around the headland. In sum, for these individuals surf-riding had become an event (see Chapter Eight), and they wanted to orient as much of their lives as possible around these encounters. In the words of Brown,

> We knew we had to live by the ocean and needed to figure out a way to make a living there
>
> (1997: 21).

When these individuals reflected upon the opportunities afforded them by existing socio-economic conditions in the 1950s, they identified two solutions to this problem[2]. As Brown goes on to state, these solutions involved either applying for employment in jobs that gave them enough time off to surf, or inventing jobs which would revolve around their surf-riding:

> Some guys became firemen (time off), some became teachers (time off).... Hobie [Alter] made surfboards, Gordon Clark made foam blanks, John Severson started *Surfer* magazine, I started making movies. Whatever we did, the main focus was how it would affect our surf time. Getting rich wasn't important. What was important was having the freedom to do what we wanted
>
> (Brown, 1997: 21).

It is important to note that unlike Blake, or the "beach bums" who made lives for themselves on the fringe of the mainstream in Hawai'i, none of this group chose to "drop out" from society, or rebel against the broader political or economic system (although Kampion and many others later broadened the surfing spaces script to explicitly relate to the counter cultural movement in the 1960s). Rather these individuals chose to find a way of grafting together their interests and economic needs. As noted above, for some this meant being fortunate enough to find jobs with favourable hours or locations (so they could surf in their spare time), but for others it involved finding ways of generating money from their surf-riding. Growing up in California (where, as Alf Anderson states, "commercialisation [is] a fundamental part" of life (2011: 128)), they actively wondered why some leisure and athletic activities had been integrated into the capitalist system (and thus become deemed "useful" and acceptable), but surf-riding had not. As Brown continues:

> Back in the early '50s, when I started surfing, the main comment from parents and non-surfing peers was "When you grow up, you'll realize you were wasting your time when you could have been doing something useful". I could never figure out why golf, tennis, baseball, football or being a cheerleader was "useful" and surfing wasn't
>
> (1997: 21).

Due to their particular cultural situation, these surf-riders saw an opportunity to attempt the integration of their own recreational activities into the capitalist system. They began to experiment with a range of "architectures" (from board-making, image-taking, and media-making, as Brown alludes to above) in order to try to make a living from surfing spaces. In sum, they tried to transform the experience of "the wave [into] a commodity to be sold to surfer audiences" (Ormrod 2005: 46).

Yet as we have seen in Chapter Seven, whenever one seeks to engage with the experience of the wave, one encounters the paradox of representation. The imagineers knew how their surf-riding cyborg identities defined them, and how the energy of the surf zone affected them, but how could they translate the momentous and momentary into something tangible and tradeable? They did so by positioning words, photos, and moving images into a broader script of surf-riding, a "central organizing idea, suggesting what is at issue" (Bennett, 1975: 65) when one engages with surfing spaces, at least from their point of view. Severson, for example, began to define what he termed the "spirit of surfing" (2014: no page), which sought to capture the moment of surfing encounter; as Severson himself states:

> One of the beauties of surfing is that it's so ephemeral. The art of the moment-fleeting. ...[what we did was] film it and paint it and strive to celebrate it
>
> (2014: no page).

and as Kampion reflects, the aim was to,

> celebrate that ephemeral almost unattainable moment, the moment we all can go, ..."oh fuck that's beautiful" you know?
>
> (Research Interview).

In translating these moments into a broader framing script, the aim was to translate something of the relational sensibilities of surf-riding to the specific youth cultures these individuals were a part of, as well as promote them to new constituencies. One way in which this was achieved was through words, but images (both still and moving) were also used. We can experience the power of the still image to capture the "spirit" of the nascent surfing script if we pause to study the early covers of Severson's *Surfer* magazine[3] (which can be found here https://www.surfer.com/cover-archive/), as well as a collection of 1950s' surf movie posters (which are reproduced here: http://www.spatialmanifesto.com/research-projects/surfing-spaces-surfer-covers-and-posters).

If one were to conduct a brief visual analysis of these re-presentations of surfing spaces, one might initially conjecture that they all focus primarily on images of the breaking wave. In them we can see the breaking wave as the product of the coming together of land, sea, and sky; we can imagine how a particular wave was rising because of a specific combination of fetch, tide, and reef. We can also plug in to the tremendous energy and motion captured by these images, the sheer "wow" that is sensed when facing them, and the involuntary intake of breath when we imagine what emotions the pictured surf-rider is experiencing in that stilled moment; as contemporary surf photographer Danny Johnson states, we can feel how these images "suck you in to try to understand what will happen, what the wave is going to do, and how the surfer is going to react" (Research Interview).

Figure 10.1 Composing Surfing Spaces.

Due to their chosen framing or artistic impression, we can identify that these images have been intentionally chosen to focus our attention on the spectacular nature of the waves and the fact that there are humans-on-boards riding them. At once the presence of the surf-riding cyborg gives scale to the image and emphasizes the size of the wave; it invites the viewer to imagine the momentum beneath the surf-rider and their skill in being able to retain balance, composure, and route in this situation. Yet although their framing enables a sense of the size of the wave, with the exception of the Orange County edition of *Surfer*, it does little to identify its location. As a consequence, these waves could be considered to be abstracted from a locale, nation, or even continent, and instead scripted as placeless and ahistorical in nature. With this in mind, it could be supposed that they have been chosen to suggest that surfing spaces are locations where it is possible to experience (a) "time out" from real life, framed as a form of utopian escape from the contemporary.

> We're a visual culture, and the [imagineers] are the cultural tastemakers. ... it's the[se] photograph[ic scripts which] are the Atlases on which our precious modern surf world is balanced
>
> (Housman, 2015: no page).

As Housman notes, images of surfing have become a vital constituent in framing the "lore" of surf-riding (after Thorne, 1976). Centripetally drawing in their audience, these images narrate a drama and beauty of often abstract littoral locations and define the characteristics required to engage with them. As Thorne states, these images are framed to emphasize the "danger", "thrill", "power", "speed", "challenge", "physical rewards" and "aesthetics" of the surf-riding experience (1976: 209), communicating the pre-literate affects of the experience in a world beyond words.

The combination of these aspects of challenge, beauty, and drama were crucial to the imagineering of surfing spaces in the 1950s. These re-presentations were successful in drawing attention to the extraordinary nature of both the littoral locations in which surf occurs, and the experience of becoming co-ingredient with them. This framing of the "extraordinary ephemeral" (after Severson and Kampion above) was in turn combined with more conventional meanings and values which resonated with the imagineers' audience. Surf images centred on young, white, male athletes, enjoying thrilling escapades in unfamiliar, spectacular moments. As a result, they explicitly sought to connect to culturally valuable ideas of sporting prowess, and celebrated a very physical, youthful, and able form of masculinity (and implied heteronormativity) to its constituencies. In this script, surf-riding was about young people, in beautiful but dangerous spaces, conducting impressive acts of athleticism all in beautiful composition; an intoxicating script was imagineered which was both radical but also conservative, combining conventional traits of the dominant mainstream, with variations of difference, cool, and even rebellion.

"Man on the wave"

> It really all boils down to the rebel, to the man on the wave, and all the rest is bullshit
> (seminal surf photographer Jeff Divine, Research Interview).

From this basis, it can be argued that the surfing spaces script as it was framed in the 1950s was also explicitly connected to broader themes which resonated in the American (and as we will explore further below, specifically Californian) imagination. As Divine pointedly alludes to above, and Lawler also identifies, there is a "individualism that resonates, somewhere, in the American psyche" (2011: 75). It can be suggested that to be an individual, self-determined and unencumbered by the constraints of society, is an essential part of the American dream, and this tenet can be seen to be directly framed into the surfing spaces script. As Severson in *Surfer* magazine writes, when faced with cultural constraints, surf-riding can become your escape:

> In this crowded world, the surfer can still seek and find the perfect day, the perfect wave, and be alone with the surf and his thoughts
> (2014: no page).

A "sense of freedom… from civilisation's complexity" (Ormrod 2005: 42) can thus be identified as central tenet to the surfing spaces script. A space was imagineered where the values, affects, and distractions of the "hustling, bustling city world of steel and concrete" (Wardy, in Ormrod, 2005: 42) could be set aside, if only for a few hours. This escape was not simply a return to "nature" in the sense of Emerson (1982), but also a return to a world without government (after Thoreau, 2018), a world in which the open frontier – so central to the American cultural imagination – was still available, if only you picked up a board and rode it; as Severson explicitly identifies:

> "Go West, young man," they said. … [And for us] the ocean is [our] frontier. …The West Coast surf [is] the end of the line. The last West is in the water
> (2014: no page).

In this way, the imagineers of surfing spaces directly aligned their script with romanticised idea(l)s of the American nation and its history. As Comer (2010) identifies, they explicitly chose to connect their pastime with a broader "W/western" narrative that resonated strongly in the national imagination[4]. As Comer explains, this surfing space script complemented the,

> classic "old western" visual and cultural tropes such as the lone white male framed against a redemptive, unpeopled landscape. He might be a mountain man or a pioneer overlooking "virgin" territories; a cowboy riding

into the cinematic distance against a backdrop of open range; or today; a solitary surfer against a seascape at sunset

(2010: 19/20).

The surfing spaces script can thus be seen to be directly associated with culturally resonant idea(l)s of the "western" frontier. More broadly, and perhaps reflecting the home location of the imagineers, further cinematic associations can be identified within their script.

> In many respects, Orange County [California] is the home of modern surfing in the same way that nearby Hollywood is the home of movies
> (Anderson, 2011: 128).

Due perhaps to the influence of nearby Hollywood, it can be identified that imagineers also attempted to infuse a degree of glamour into their surfing spaces script. As surf-journalist Chas Smith puts it:

> There is something magical about [the] surfing [spaces script]. Dreamy, floaty, sexy, fairytale-ish. Magical. Sliding across the ocean powered by nature herself. Getting a gorgeous tan. Delicious babes on the sand. ... [it was because of this script that] when I surfed, I felt California cool
> (2013: 3).

Smith even suggests that due to the "dreamy, fairytale-ish" framing of the script, its leading players were elevated into the role of cinematic heroes, handsome and virile, a divine set of humans that had been "touched" by the gods:

> ... [In this script] surfers were [depicted as] so divinely handsome. Touched by the finger of Venus. ...Surfing [in this view] is an activity where young, bronzed boys filled with health and vitality strip virtually nude and go sit in warm seas. They are often blonde. They are often handsome. ...It is a [set of] totally aspirational element[s which make] a beautifully aspirational life
> (2013: 20).

By combining the extraordinary with the radical and the familiar with the filmic, surfing spaces were scripted through a frame of novel adventure, conventional American idea(l)s, and quasi-mythological aspirations. By incorporating cinematic allusions with more "classic" tropes, the script had the effect of "transport[ing its constituency's] imagination to a level beyond the factual" (after More, 1996: 234), elevating ordinary lives through an aspirational imagineering. As we have seen, this script was developed through images and words in surf magazines, but it was consolidated and disseminated through surf films; and no example is more definitive of this process than the seminal surf movie, *The Endless Summer*.

Consolidating the surfing spaces script: *The Endless Summer*

Reflecting the aspirational and escapist nature of Hollywood cinema, films played a key role in developing and disseminating the surfing spaces script. Although *The Endless Summer* was not one of the earliest surf films, or even the first "surfari" film, it nevertheless made a profound contribution to the imagineering of surfing spaces. As Comer suggests,

> it is not too much to say that *The Endless Summer* produced both the initial economy and the foundational structures of feeling that today underwrite surfing as an international public culture
>
> (2010: 23).

The Endless Summer told the story of two young, male surfers travelling the world in search of waves. As such, this movie helped to define the surfing spaces script as much about exploration as it was about the surf-riding experience. *The Endless Summer* provided the template for surf-riders to travel in their quest for stoke, extending the aspirational moment of wave-riding into a broader lifestyle activity which involved global travel to exotic littoral locations across the globe. As a consequence of the success of *The Endless Summer*, "travel and adventure [became] the new frontiers of surf in the sixties" (Severson, 2014: no page), and the surfing road trip came to be defined as "one of the fundamental truths of your surfing life" (Evans, 2020).

Pitched towards the growing youth markets of America (which the director Bruce Brown understood through hundreds of college voiceovers for his previous surf movies), *The Endless Summer* documented a world of surfing spaces which, up to that point, few American surf-riders had knowledge of. Aspiring to be a personalised document of encounter rather than a rigorous documentary, the film epitomised a common tone of the surfing spaces script, combining travelogue with frat-house ribaldry, inviting its audience to empathise with the young, white, male surfers who sought to explore "new" frontiers and discover "new" waves. The film did not intend to broaden the cultural horizons of its constituencies or challenge the nationalist imaginaries which underpinned their sense of self (and the nascent script of surfing of which it was a part). Rather it sought to consolidate the uninformed prejudices of its audience by unreflexively celebrating the "discovery" of these new waves and perpetuating the notion of the world as an open frontier. In short, it continued the broader praxis of the colonisation of surfing spaces witnessed in Hawai'i which instigated the "misleading narrative of [American] surfing 'pioneers' on a mission to uncover seemingly ... unknown breaks" (Warren and Gibson, 2014: 35).

Indeed, the framing of the places visited by surfers in *The Endless Summer* had the effect of making explicit the colonial imaginary that defined the praxis employed by Hume in Hawai'i and was to become intrinsic to the surfing spaces script. The local people in the destinations visited were cast as ignorant of surfing technologies and practices, as Thomson suggests, the film, assumed they [the American surf-riders] introduced surfing to Africa, proclaiming [their arrival was] the "start of bellyboard surfing in Ghana." [However] this was not the start of surfing [in that area], but rather a display of old traditions. Indeed, several kids had round-nosed body boards, while their surfing fluencies indicate that they were not neophytes

(2017: 141).

The privilege and entitlement of the travellers defined the narrative, as "classic colonial stereotypes of African cannibalism, timeless village traditions, mask-wearing jungle men, and so forth, reign[ed] unchallenged" (Comer, 2010: 63). Indeed, as Comer goes on: "Besides the sophomoric racisms of the narrative, the film simply registers no sense of the history of colonial presence in Africa" (ibid) and reflecting the cultural amnesia over the "century of tears" (see Chapter Nine), according to Comer the film, "clearly had no concept of Hawai'i as a colonised place, nor of the 'surfers themselves as complicated or unwitting agents of colonisation" (2010: 61).

The Endless Summer thus scripted surfing spaces as part of "a colonial dreamscape" (Laderman, 2014: 55), a place "beyond the factual" (after More, 1996: 234) where young, white, middle class America could embark on a hedonist search for freedom from the restrictions of their own cultural (b)orders, regardless of the consequences (which we will explored in Chapter Sixteen). Here were heroes scripted in the national image, who were "gonna explore the world, our way" (Pat O'Connell, star of *Endless Summer II*, 2020); for them, non-US cultures were simply open frontiers waiting to be taken and made by the values of the west. This "exceedingly simplistic surfing imagination..." (after Laderman, 2014: 4) intentionally perpetuated historical processes of empire and colonisation, "othering" whole populations in order to entertain the privileged[5].

Imagineering the surfer as western archetype

As we have seen, the surfing spaces script as imagineered in the 1950s and 60s initially emanated from a wish to capture the "fascination with heroism, peril and the transient and ferocious nature" of the littoral (after Gartside, 2020). Surf-riding was then associated with broader cultural idea(l)s, framing it as a cultural geography of discovery, encounter, and self-determination which would attract an audience of white male youths in America. The script offered an adrenalin shot of escape and rebellion, whilst tying them into traditional cultural tropes of the frontier and the west. In this way, the script was not a faithful rendition of reality, but a set of symbolic framings which gave values and meanings to surf-riding which would appeal "on a rational and emotional level" with the target culture (Nijd and Peters, 2002, cited in Kuiper and Smith, 2014: 8, see above). The fact that this script actively avoided any "confrontation with the realities of political economy and the circumstances of global power" of which it was a part (after Routledge, 1997: 360) was not a problem, indeed was part of its appeal. The escapist nature of the narrative was vital, as Severson acknowledges, the imagineers, "took liberties right from the beginning, realizing that [in order to be successful, it needed to be] more fantasy than [a] document... of fact" (2014: no page). Existing in a realm "beyond the factual" (after More, above), the script was accepted as an "illusion" (after Hening, 2015: 24), but an incredibly powerful one. Both for the imagineers and their core audience, the script offered "a thrill-ride of control" (after Storr, 2019: 98); enabling them to (b)order their world in line not only with their own privileges, but also their own priorities.

The surfing spaces script was thus imagineered as part-real and part-fantasy, connecting embodied practice directly to the idea(l)s and mythology of the western world. Indeed, this chapter concludes by arguing that the surfing spaces script gained popularity and prominence because these traits can be seen to directly replicate the most successful symbolic "archetype" used in the majority of the "story-lines" and "frames" (after Gamson and Wolfsfeld, above) of western culture. Fiedler (1966) and Conard (1975) outline what they describe as key tenets of this dominant archetype. Firstly, this dominant myth positions a male protagonist as central to the story, and by extension is generally defined by an absence of strong female characters. Secondly, the narrative focuses on the protagonist's adventures beyond mainstream society as he seeks to escape and define himself "in isolation". Thirdly, water plays a key material and symbolic role for the protagonist (as "teacher of and initiator into the mysteries of life" as well as being indicative of "life and rebirth" (after Conard, 1975: 359)); and fourthly, the male protagonist enjoys the subservient support of male companions who are often caricatured as "the pure, natural man, unintellectual and close to nature" who "represent a lower social order than the hero" (ibid.). These conditions, Conard argues, define the "working parameters" for "isolato-archetype in Western literature" (1975: 359).

This archetype can be seen to resonate strongly with surfing spaces as scripted by imagineers in the 1950s and 60s. Here, surf-riders are framed as male (and by implication heterosexual), who seek self-determination through escape from mainstream society. This is realised in practice through adventure and discovery in the (global) littoral zone, a space which enables individuals to re-invent their identities through the extraordinary activity of surf-riding. As they do so, this script encourages them to caricature potentially problematic cultures and politics, relegating any and all pre-existing local populations to subservient and supportive roles to the heroes of the narrative. Harnessing the traction of this dominant archetype, the surfing space script provides actual and potential surf-riders with their own "mythology", scaffolding their practices within a framework of values and aspirations which give "the sense of living in a meaningful story" (after More, 1996: 238). This script invites the youth of America to sense an escape from the limits of their cultural horizons whilst maintaining the privileges they enjoy as part of that culture. To paraphrase the author Aldous Huxley[6], it offers them the sense that they could become "somebody", simply through surf-riding; it identifies that: "…a man [wants to be] somebody… At home, he [may be] lost in the crowd, he does not count, he is nobody", and in response, offers him the opportunity to

> satisfy the profoundest and most powerful of all the instincts – that of self-assertion. …No wonder … he loves the [script, it is] a kind of intoxicant. … the sense of power which it gives, the feeling of grandeur and importance are …not *entirely* an illusion. …[Why not be a surfer?] What man likes to be sediment, when he might float gallantly on the sunlit surface [of the sea]?
> (Huxley, 1957: 12, my emphasis).

Conclusion

The imagineering process thus transformed the energy of the littoral zone into a script which changed how the moment of surf-riding, the individuals involved in the process, and the locations in which it occurs, were popularly understood. Surf-riding was no longer simply about the moment of being co-constituted through an encounter with land, sea, and sky, it became a mythic move to escape the (b)orders of the mainstream, an opportunity to assert your individuality in a new frontier. Surfing was scripted therefore not as a simple form of recreation but as a tool of cultural identity, with complex consequences for the positioning of surf-riders in respect to broader processes of colonisation and globalisation. In Chapter Eleven we move on to explore the consequences of this script in detail and how its symbolic framings and media architecture enabled it to spread on an international scale. We also begin to chart the changes it brought to how the cultural geographies of surfing spaces have been understood, marketed, and practiced since its composition.

Notes

1 The use of the male pronoun here is intentional, and reflects the particular masculine bias of the surfing spaces script as it was initially imagineered, and subsequently perpetuated into the twenty-first century (and this bias will further explored in Chapter Fifteen).
2 It is important to acknowledge that these individuals were able to engage with solutions to their problem due to their relatively privileged status (as white, middle class, male citizens) during this period of American history.
3 *Surfer* was created by Severson from a small one-off magazine in 1959–60 to support his *Surf Fever* movie, and ran in print until 2020.
4 Indeed, at the time of its scripting, this direct allusion to the "west" and the open frontier was actively being harnessed in more mainstream political discourse. As Dominguez Andersen notes, President Kennedy harnessed the "influential symbolism of a New Frontier… [in] an attempt to evoke and rearticulate the rugged, individualist virility of the nineteenth century's westward expansion" (2015: 516). As he notes, "In 1960, John F. Kennedy announced the newly proclaimed state… of Hawaii …to be "[a] "symbol… of the new frontier" and linked [it] to America's national and global future: "The spirit of adventure that drove men to … Hawaii

must be awakened again in the American people if we are to maintain our position as a leader in the free world" (ibid.).
5 Such positioning may have risked critical ridicule on the movie's release. However, as Comer states: "...No one took offense at the film's unselfconscious coloniality, its global confidence, or its unconcerned ignorance" (2010: 63). Indeed, within surf- and mainstream popular -culture, *The Endless Summer* was heralded as, "creating and defining an entire category of cinema" (Surferpedia, no date). In 2002, it was selected by the Library of Congress for preservation in the United States National Film Registry for its "cultural, historical, or aesthetic significance" (ibid.) and, as Warshaw puts it (in the *Surfer*'s History of Surfing Films, "For the first time the rest of the world would have a clear look at the surfing lifestyle.... *The Endless Summer* defined our sport" (cited in Ormrod, 2005: 48).
6 Originally written with respect to colonial English males experiencing India.

11 The International Influence of the Surfing Spaces Script

Introduction

As we have seen in Chapter Ten, in the 1950s and 60s surfing spaces underwent a process of imagineering. Drawing on their own experience, as well as complementing the marketing of surfing by Hume and the cultural creativity of Blake, a small number of surf-riders in California transferred the "stoke" of surfing spaces into a "script". Using a range of media architecture (including words, posters, magazines, and films), this script gave meaning to the key characteristics of surfing spaces as these surf-riders viewed them. This script created a new version of surfing space which resonated strongly not only with the contemporary context, but also the development of western culture since that time. The script was so successful that surfing's "big bang" occurred: surfing went pop.

Rather than re-counting the successful growth of the surfing spaces script since the 1960s (for this see Kampion and Brown, 1997; Colborn et al, 2002; Fordham, 2008; Warshaw, 2010; Comer, 2010; Lawler, 2011; Westwick and Neushel, 2013; Laderman, 2014; and Hough-Snee and Eastman, 2017 as examples), this chapter seeks to emphasise the role of the imagineers in creating this success story. It argues that the ability of the imagineers to connect and control disparate aspects of their media architecture not only enabled them to communicate a coherent and mutually beneficial script to their audiences, but also extend the territories over which their script could have effect. As the script extended from America to overseas, it not only demonstrated the power of a well-constructed and disseminated meta-narrative to infuse meaning and aspiration into a recreational activity, but also the possibility for this to be a lucrative endeavour. This chapter argues, therefore, that not only did this imagineering process offer a script to those involved in surf-riding that confirmed this practice as more than a simple pastime, it also confirmed that it was possible to occupy a new position as a "surf-lifestyle entrepreneur", in other words, to make a living from moulding and shaping the codes associated with the surfing spaces script. This chapter suggests that the imagineering process thus created the basis for a new culture industry, including not only a market of people who wanted to buy into this script, but also a new profession of "surf-lifestyle entrepreneur" whose role was to serve and influence them. The

DOI: 10.4324/9781315725673-11

imagineering of this script, therefore, was central to the formation of the cultural geographies of surfing spaces today, with all their possibilities, paradoxes, and problems. From this basis, the cultural "codes" (see Chapter Ten) which implemented the surfing spaces script in practice were generated, maintained, and reproduced. The conviviality, conflict, power, and politics which these codes produced will be explored in the remaining chapters of Part II.

Scripting success

As we have seen in Chapter Ten, the imagineering of a surfing spaces script in the 1950s and 60s was a working solution to a practical problem faced by a loose coalition of relatively young Californian surf-riders. From craft-development to image-taking, from clothing-design to media-making, these individuals created a range of "architectural" mechanisms to integrate surfing spaces more fully into the capitalist system. These individuals thus not simply rode the waves they also composed the script and controlled its dissemination, employing a range of media including films (as we have seen in Chapter Ten), merchandise (see below), and magazines to control the nature of surfing spaces.

For example, following the success of small pamphlets circulated at film screenings, Severson produced *The Surfer* magazine in 1960, which sold 5000 copies within days (see Warshaw, 2010), soon evolved into a quarterly, and from 1960 to 1970 its per-issue circulation expanded to approximately 100,000 (Robertson, 2020). *Surfer*, alongside other magazines including *Surfing International* (which evolved into *Surfing*), celebrated and disseminated the newly-formed script, making stars out of particular surf-riders and promoting specific styles of surf-riding (as we will see in Chapter Fourteen). As American surfer, author, and journalist William Finnegan reflects, at this time,

> Magazines, particularly *Surfer*, had become the main conduit of the surf subculture's self-celebration, and [we] read the mags avidly and talked, with increasing authority, in the new language [we] found there
> (2015: 70).

As Gamson and Wolfsfeld suggest, there is often a "competitive symbiosis" between media architecture and particular cultural groups (1993: 116)[1]. However, in the case of the imagineering of surfing spaces, this symbiosis was a creative one. As the imagineers were part of the culture they were actively shaping, their personal experience and authority lent credibility to the script and their media outlets, and in turn, this validated and affirmed those consuming it, encouraging their practices and celebrating their surf-riding endeavours. This symbiosis was also commercially successful too. Not only did the credibility of the media drive its circulation, but as the imagineers were informally linked through their shared cultural involvement, there were also opportunities to be realised through the close connections between cultural celebration and advertising (see also Grannis, 2013). As Wolfe states, there [was] a kind of back-and-forth thing [between the culture and the media]. ...Bruce Brown will do one of those incredible surfing movies and he is out in the surf himself filming Phil Edwards coming down a 20-footer in Hawaii, and Phil has on a pair of nylon swimming trunks, which he has had made in Hawaii, because they dry out fast – and it is like a grapevine. Everybody's got to have a pair of nylon swimming trunks, and then the manufacturers move in, and everybody's making nylon swimming trunks, boxer trunk style, and pretty soon every kid in Utica, N.Y., is buying a pair of them, with the competition stripe and the whole thing, and they never heard of Phil Edwards. So it work[ed] back and forth

(2002: 111).

As this example suggests, due to the imagineers combining cultural and media roles, they obtained a high degree of power and authority over the script of surfing spaces, and in turn over surfing spaces themselves. As the imagineers infused meaning and aspiration into their script of surf-riding, their innovations in craft and clothing could be directly communicated to their growing constituencies, with the media architecture driving the process; as Kampion and Brown state, movies and "surf magazines tie[d] it all together" (1997: 93).

As the media architecture of the twentieth century transgressed regional and national (b)orders, the ideas and values disseminated by these Californian imagineers also had the capacity to go global. The potential market[ing] opportunity for this script of surfing spaces was thus of a different magnitude, speed, and scale to that in the Hume or Blake eras, through local consumption of Californian surf media it was now possible that,

Surfers in Australia or San Diego or Cape Hattera could now see what surfers in Hawai'i or Malibu were doing. They could see what kinds of boards they were riding, the maneuvers, the nuances, who was hot and why, and – above all – what was possible!

(Kampion and Brown, 1997: 93)[2].

Just as youth cultures in America were seduced by this script of surfing spaces (see Ormrod, 2007; Lawler, 2011), young people in many other countries felt not only the powerful "hum" of the waves themselves (Hulet, 2011: no page), but also the "California cool" of the imagineers' framing (after Smith, 2013, see Chapter Ten). As British surf-rider and -historian Roger Mansfield states (with reference to endogenous surfers in the UK), as the script spread "people lapped it up" (2009: 172). In the UK, for example, where from "the 1920s... it was quite usual to see people in the summer using small planks in order to play (prone) with the white water (bellyboarding) at the beach" (Esparza, 2016: 199)[3], the arrival of the surfing spaces script strongly influenced the development of endogenous practices. As Mansfield explains,

> To the innocent early British surfer, [*Surfer*] magazine provided answers about everything – waves, boards, style and attitude
>
> (2009: 180).

These media architectures thus set the new framework for existing surf-riding not only in different parts of America, but in overseas locations too. It positioned their activities within a wider context, a (recuperated) history, and set of aspirations that appealed not only to local surf-riders, but also attracted new participants. As Mansfield goes on to state, in the UK for example,

> Young people across the country were inspired to try surfing when *The Endless Summer* went on general release in cinemas in 1964. ...It reached out to ordinary young guys and generated a positive image that reverberates to this day. The surfer became a symbol of health, adventure and youth
>
> (2009: 172).

This dissemination of the surfing spaces script went hand-in-hand with the spread of popular culture too. As Hollywood movies and the music industry sensed the opportunity that youth cultures and the surfing spaces script now offered (see for example Kampion and Brown, 1997; Fordham, 2008; Warshaw, 2010), the attraction of this new way of living was clear; as Mansfield explains,

> As surfing was taking off in California, a new style of music emerged from the West Coast – the reverberating "surf guitar" sound of bands like Dick Dale & The Deltones. Not long after, vocal harmony groups like The Beach Boys and Jan & Dean became hugely popular around the world. They communicated lyrically what they felt about sunny Southern California youth culture. Their statement was simple and relevant to the time: cars, waves and girls were happening. It all promised fun, fun, fun... Post-war austerity was over...
>
> (Mansfield, 2009: 172).

The surfing spaces script which emanated from the imagineers in California thus spread internationally through its media architecture. As it encountered new or existing local surfing space assemblages, the authority and aspiration of its archetype meant it could "provide the answers" about what surfing spaces could and should be like. Although the dominant script was often adapted, mutated, or cross pollinated with local cultural geographies (see Booth, 2001; Ormrod, 2007; and Thomson, 2017 for examples)[4], across the years that followed the constancy of its values and idea(l)s generally remained, and as Lawler sums up, "in its every incarnation, the content of the surfer image has remained largely consistent" (2011: 2).

Embodying the American Dream

Due to the creative and commercial symbiosis of their roles as both practitioners of the culture and its media, imagineers were in a unique position to validate their interpretation of what surfing spaces should be. As we have seen in Chapter Nine, due to the relatively exclusive cadre of people who imagineered this process, it was possible for editors and journalists to ensure synergy between media and script, ensuring that "certain actors were given standing more readily than others" and if necessary, any discrepant actors, "rendered invisible" (after Gamson and Wolfsfeld, 1993: 119). In this way, the media architecture put in place by the imagineers of California "dominated" the scripting of surfing spaces through their capacity to control their message and reach "wide audiences" (after Surfer Today, 2013). Due to this control, it can be argued that the script became naturalised for its audience, with its "social constructions rarely appear[ing] as such to the reader" (after Gamson and Wolfsfeld, 1993: 119). The distinction between surfing culture and the script itself had become elided to a large degree, with the script normalised into a "transparent description of reality, not as [an] interpretations" of it (ibid.). The filmmakers, photographers, contributors, and editors of these architectures thus consolidated their position as the "opinion-makers" of their cultural geographies (Surfer Today, 2013). From this time onwards, as Stranger confirms, this architecture "play[ed] a key role in the mediation of surfing's collective consciousness through the creation and dissemination of images and meanings, myths and legends" (Stranger, 2011: 188). With reference to print media specifically, these early publications "created and defined the surf magazine and, in doing so, an industry and a good measure of the sport itself" (Kampion and Brown, 1997: 93).

These imagineers could thus be argued to living embodiments of the American dream: they opened up their own new frontier, became masters of their own destiny, and made money doing it. As Wolfe states:

> [Brown] goes out on a surf board with a camera encased in a plastic shell and takes his own movies and edits them himself and goes around showing them himself and narrating them at places like the Santa Monica Civic Auditorium, where 24,000 came in eight days once, at $1.50 a person, and all he has to pay is for developing the film and hiring the hall. John Severson has the big surfing magazine, *Surfer*. Hobie Alter is the biggest surfboard manufacturer, all hand-made boards. He made 5,000 boards in 1965 at $140 a board. He also designed the "Hobie" skateboards and gets 25 cents for every one sold
>
> (2002: 111).

These activities were so successful that, as Wolfe goes on to explain:

> Bruce Brown and [John] Severson and [Hobie] Alter are [now] known as the "surfing millionaires". They are not millionaires, actually, but they must be among the top businessmen south of Los Angeles. Brown grossed something around $500,000 in 1965 even before his movie *Endless Summer* became a hit nationally. [Alter] grossed between $900,000 and $1 million in 1964
>
> (2002: 111).

These imagineers were therefore successful in their original aim, to create a lifestyle for themselves which could balance their commitment to surf-riding with the ability to make money from it. In the context of 1950s and 1960s USA, this success was itself a form of heroism for many young people who wished to feel a sense of escape from the privileges of middle-class conformity and suburbia; as Lawler puts it, for these young people,

> the only adults they look up to are guys like Bruce Brown, John Severson, Phil Edwards, and Hobie Alter, thirty-something surfer/entrepreneurs who are exempt from the criticism accorded other adults because they never "haired out" and gave up *the life*
>
> (2001: 110).

This imagineering process was as motivating and aspirational as the surfing spaces script itself, as Wolfe states:

> God, if only everybody could grow up like these guys and know that crossing the horror dividing line, 25 years old, won't be the end of everything. One means, keep on living The Life and not get sucked into the ticky-tacky life with some insurance salesman sitting forward in your stuffed chair on your wall-to-wall telling you that life is like a football game and you sit there and take that stuff. The hell with that! Bruce Brown has the money and The Life. He has a great house on a cliff about 60 feet above the beach at Dana Point. He is married and has two children, but it is not that hubby-mommy you're-breaking-my-gourd scene. His office is only two blocks from his house and he doesn't even have to go on the streets to get there. He gets on his Triumph scrambling motorcycle and cuts straight across a couple of vacant lots and one can see him... bounding to work over the vacant lots
>
> (2002: 111).

The imagineers of surfing spaces were thus successful in aligning three mesmeric idea(l)s of the American cultural imaginary: freedom, self-determination, and profit. Their activities offered much to middle-class, white America, who despite benefitting to a large degree from a growing economy, employment opportunity, and general prosperity, nevertheless felt a degree of relative deprivation in terms of adventure and excitement[5]. The example of these imagineers therefore suggested to young people that they could create opportunities for themselves that were different to those offered to their parents, and reinvent their own spaces in line with their own eventful practices.

A new breed of "insider"

As we have seen in Chapter Nine, Hume colonised Hawai'ian-Californian surfing spaces through his particular version of commodification, and subsequently Blake employed his own cultural creativity to further influence the nature of these spaces. In the 1950s and 60s, however, the imagineering process can be seen to propagate a new breed of individuals who scripted surfing spaces to re-organise its "cultural, financial, social and human capital… to generate revenue" for themselves (after Anheier and Isar, 2008: 1).

Through these processes, this chapter argues that these individuals should be understood as a new version of cultural entrepreneurs (see also Chapter Nine, and Lawler, above). In contrast to marketeers like Hume, these individuals imagineered a culture they were involved in, assuming the entitlement to commodify this culture because, in effect, they were commodifying an idealised, recuperated, and mythologised version of themselves. Capitalism was therefore not an afterthought to their lifestyle (as it could be argued it was for individuals like Blake), it was intrinsic to it. In this case, capitalist "others" did not "steal" these surfing spaces (and all they got was a baseball cap, see Taylor, no date), rather these "insiders" (see Maloney, et al, but also Stranger, 2011), exploited their own first-hand understanding and expertise of surf-riding in the Hawai'ian-American surfing space and imagineered it into a commodity they could sell. As a result of this positionality, these individuals can be understood as the germinal form of "lifestyle(-sport)" entrepreneurs (see Wallis et al, 2020).

As Wallis et al (2020) identify, lifestyle entrepreneurs occupy a difficult to define middle ground in relation to the often-dualised positions of work and play. Reflecting perhaps the liminal position discussed in relation to the littoral zone in Chapter Three, and third space positionality in Chapter Two, lifestyle entrepreneurs occupy combined positions (after Jones et al, 2017); they share similarities with hobbyists on one hand, and capitalists on the other. As such, they are not wholly defined by either label, but exist with aspects of both in their practices. As a lifestyle entrepreneur, individuals seek their own personally-defined balance between the apparently opposing objectives of profit and pastime. From commodifying themselves they generate money to live on, whilst also wishing to save enough time to play in. As Brown put it in relation to his own entrepreneurial activity:

> What was important was having the freedom to do what we wanted. It didn't mean we didn't take our jobs or professions seriously; we did. It meant keeping things in perspective. …people in the movie business would tell me, if I wanted to be successful in the motion picture industry, I would have to move to Hollywood. I said I would rather be a milkman at the beach than live in Hollywood. I wasn't kidding then, and I'm still not kidding now
>
> (in Kampion and Brown, 1997: 21).

As such, profit would not be the only "bottom line" in these imagineers' activities. (Surf-) lifestyle entrepreneurs had to "bridge" the opposing identities of "creative" and "business- person" (Eikhof and Haunschild, 2006), attempting to balance profitability with being "true" to their pastime's orientation (after Bredvold and Skålén, 2016). The imagineers of surfing space had to juggle these objectives, as Warshaw states with reference to Severson, his,

> was an impressive balancing act. There were checks to deposit, meetings to chair, advertisers to court. Severson did all that. There were also waves to discover and ride, and a drive to present the whole experience to his audience not only through journalism but art. Severson did that too. Yes, he wanted readers to go out and buy the products advertised on the magazine's pages. He also wanted to remind them… that what they were doing was beautiful and unique
>
> (2010: 177).

Although "insider" status helped to transfer a degree of authority to their new script (as Stranger also argues in relation to surf companies in Australia in the late twentieth- and early twenty-first century (2011)), their dual positioning as lifestyle entrepreneurs meant in practice they welcomed capitalism into the heart of the culture. Although it was not the case that the culture was immune to or outside of capitalism up until that point (as we have seen in Chapter Nine), nevertheless the direct connection between capitalism and culture was not a neutral process. Indeed, the "enmeshment" of capitalism and culture (after Evers, 2019: 423) meant that a core meaning of the imagineered script was "distorted" (following Hough-Snee and Eastman, 2017, and Gartside, 2020, see Chapter Nine) from the outset. As we have seen in Chapters One and Nine, to be worthy of the name, surfing is scripted to be pure fun. On this basis, many endogenous practices have been sidelined or ignored as illegitimate by these imagineers due to their association with fishing or other maritime industries. This has enabled the Californian script to write itself as the definitive version of surfing space. However, as we have seen, imagineering is itself a form of industrial entrepreneurship, intrinsically defined by its hybridisation of commodification and/of pastime. An irony at heart of the surfing spaces script is thus generated: those who imagineer the cultural geographies of surf-riding fall beyond their own definition of what authentic surf-riders must be. They therefore become a problematic band of professional "insiders" (perhaps better termed "thresholders", after Maloney et al, 1994) who despite their power to define surfing spaces cannot, by the terms of their own script, be included as authentic surf-riders as they have chosen to sell their stoke. Thus imagineering was not a neutral process, it created a commodity where there wasn't one (or many) before, it put a price on a value, and promoted it to all whilst it celebrated the few. Severson was sensitive to the tension at the heart of the imagineering operation, explaining that:

> I've always leaned toward celebration, rather than promotion. I felt we were talking directly to fellow surfers, rather than sending messages to attract new surfers. … I considered [this tension] when I started *Surfer*, but it was a big factor by the time I left, and was perhaps part of my reason for

leaving. [Yet it's true that] anyone in the surf-related businesses or the media is faced with being a part of expanding surfing, but many of us came out of the Depression, and paying the bills to stay close to surfing was our goal

(2014: no page).

Thus the imagineer's position as a surf-lifestyle entrepreneur complicated their relation to, and the definition of, the surfing spaces script. In one sense, they were creative, yet in another they were contaminating, commodifying their practice whilst selling the dream of an escape.

Conclusion

We can see therefore that the imagineering of the surfing spaces script can be seen to introduce the role of the "surf-lifestyle entrepreneur". From contingent solutions to local problems, individuals like Severson were successful in imagineering, "surf language, surf music, surf art, surf media, surf fashion – all the basic elements of what are now considered essential to modern surf culture" (Barilotti, 2013: 16). Through so doing, they created the possibility for the ""surf industry", "surfing media", "surfing industrial complex", and "surfing culture industry"" that followed (after Hough-Snee and Eastman, 2017: 85). As Severson himself puts it, when asked the question, "What do you think your role has been in surfing?", confirms that

If you're working in the surf industry [today], you might consider that I helped make that possible

(2014: no page)[6]

The imagineering process and the script it produced resonated strongly with many young people's desire to acquire a sense of freedom and autonomy. That this freedom may have been symbolic, and easily acquired through purchase perhaps made it even more popular, as a sense of rebellion could be achieved as easily as attending a drive thru movie. However, as Smith suggests (2013, and Chapter Ten), the surfing spaces script nevertheless offered a sense of "California cool", a gateway drug which seduced many into this version of the cultural geographies of surfing, and as a consequence, encouraged more to get involved in the practice. Thus, both then and now, these entrepreneurs not only made and continue to make money, they also shaped and continue to mould the nature of surfing spaces. Even though, in the words of Horton, "each component of [the] architecture might by itself seem rather insignificant" when taken together they can, "result in the production of a coherent lifestyle" (2003: 64)[7]. In this way, from the 1950s onwards photographers, writers, editors, and advertisers combined to, "control the clothes and shorts that we [i.e. surf-riders] wear, and even the way we ride waves" (Surfer Today, 2013). As West puts it, these imagineers became, so important that definitions of who is and who is not part of the surfing subculture now depends not on who surfs but on who is portrayed and represented in the specialist surf media. As Leanne Stedman said, "It can now be argued that the subculture as it is simulated through magazines and films *is* the surfing subculture" (1997, 78)

(2014: 411).

The imagineers were thus successful in framing a powerful surfing spaces script, one which was so well disseminated and became so popular that it had the potential to control surfing spaces themselves. Yet, as we will see in the following chapters, scripts and architecture cannot influence geographies on their own. To be successful they need to work in combination with cultural "codes". The success of the imagineering process gave those involved in it the power to prescribe and impose these codes. Following the development of the script, these imagineers (and their successors in the surf media and broader surfing industry) have enjoyed the capacity to "simultaneously pander to surfing's fabled creativity" whilst they also "rigidly define boundaries of surfing activity" (Hough-Snee and Eastman, 2017: 85), in other words, they have enjoyed the ability to "strongly ascribe" (after Melucci, 2006) the nature of authentic surf-riding "performances" (after Horton, 2003). In Chapter Twelve we introduce the idea of cultural "codes" and investigate not only the (b)orders of access and entitlement they have created, but also the inequalities and marginalisations they have imposed as they translate the script of surfing spaces into practice.

Notes

1 As Gamson and Wolfsfeld put it, "most conversations between [cultures of practice and cultures of media] take a drearily predictable form: 'Send my message,' say the [cultural actors]; 'Make me news,' say the journalists. In this dialogue of the deaf, neither [party] make an effort to understand how the other views their relationship or, better yet, the complex nature of these transactions" (1993: 115). In relation to the imagineering of surfing spaces, this complex relationship was simplified as those who controlled the media also controlled the meaning, and to some extent at least, provided the merchandise which propagated the surfing space script, ensuring creative and commercial symbiosis for the actors involved.
2 The presence of surf-riders in these locations may have been a direct influence of Hume's promotion or Blake's innovation, but it is also possible that these were endogenous surfing spaces which would soon become influenced by the script of surf-riding as imagineered in California.
3 As Esparza argues, "The first image of actual surfing (stand-up) in Britain is connected with a private film by Lewis Rosenberg from 1929. This movie was 'discovered' for the surfing community in 2011. Rosenberg shaped a surfboard on his own from balsa wood, inspired by a documentary about Australia, where surfing images appeared" (2016: 199).
4 Drawing on the work of Booth (2001), Ormrod (2007: 89) states, for example, that "American surfing narratives [as] represented in surfing magazines and 'pure' surf films" were often adapted, "dependent upon national agendas". Ormrod suggests, for example, that "a 'typically' British indigenisation would be the 'stoic adventurer'"

who emphasises "the hardships as well as the pleasures" of surfing spaces (see also Chapter One). Further national adaptions may include the Australian "surfer-as-warrior" (Evers, 2010) or "surfer-as -rebelliously-adolescent" (see Henderson, 2001).

5 Capitalising on this imagineering process was inspiring for those beyond America too, as Mansfield states:

"…reading *Surfer* magazine and watching the occasional surf film, young [British] surfers … adapted their unique identity to fit in with the simple reality of making a living. The artisan became the surfboard builder or wetsuit maker. Other surfers were channelled into competition, while others capitalised on the growing commercial niche in shops and surf fashion companies" (2009: 10).

6 Contemporary surf editor and journalist Sam George concurs, suggesting that:

"Before John Severson, there was no "surf media", no "surf industry" and no "surf culture"" (in Weisberg, 2020: no page),

Whilst Kampion and Brown acknowledge it was John Severson, "who started it all" (1997: no page).

7 Horton who was writing in relation environmental cultures (see Horton, 2003).

12 Coding Surfing Spaces
Surf-rider Positioning

Introduction

As we have seen, in the twentieth century a dominant version of the surfing space assemblage emerged. In this assemblage, the experience of wave-riding was transformed into a "script" infusing surfing spaces with ideas of freedom, autonomy, and individualism. This script resonated strongly with western archetypes of open frontiers and self-determination, and successfully employed a range of media "architecture" to disseminate its values throughout Western America, and then to key surfing markets around the world.

Yet as we will see in the remainder of Part II, scripts and architecture cannot influence geographies on their own, to be successful they need to work in combination with cultural "codes". Codes are (b)orders of practice which define what any cultural geography could and should be. As we have seen in Chapter Eleven, in relation to surfing spaces, the success of the imagineering process meant that those involved had the dominant authority to define, maintain, and reproduce these codes. As we will see, these cultural codes may detail *how* to surf-ride in an acceptable way, *who* is defined as a proper surf-rider, and *what* technologies or craft should be employed to surf authentically. The imposition, adoption, and subsequent perpetuation of cultural codes complete the imagineering process, and prescribe (b)orders of access, entitlement, and marginalisation. This book continues by examining and critiquing the codes which constitute the dominant assemblage of surfing spaces. It does so through identifying two key orientations around which these codes are based: firstly, the geographical *positioning* of the surf-rider on the wave, and secondly, the cultural *positionality* of the surf-rider. Chapter Twelve proceeds to detail the importance of geographical *positioning* on the wave (and the way this positioning in turn valorises a particular version of the surf-rider, and particular styles of surf-riding). The chapters that follow focus on the *positionality* of the surf-rider specifically, and the ways in which cultural ideas of socio-demographic eligibility have leached into surf-riding codes. Each of these chapters demonstrate the rise to prominence of these surfing space codes, their contradictions with the overarching script, and how each code is being actively challenged in contemporary practice. Taken together

DOI: 10.4324/9781315725673-12

these contradictions and challenges are significant, and what they ultimately mean for the sustainability of the dominant cultural codes, and by extension the power of the surfing spaces script and its imagineers, will be considered in Chapter Seventeen.

Coding culture

> ... we have this idealised image of the soul surfer on his or her own roaming the planet without anyone else around. But as soon as you've got more than one person on a beach, you've got a culture...
> (Martin, British surfer and academic, in Capp, 2004: 221).

> Like all surfers I feel the weight of discipline imposed by surfing culture
> (Booth, 2008: 17).

As we have seen in Chapter One, littoral locations can be "spatialized as margins, as "free zones" (Shields, 2004: 44); as scholar and surfer Evers puts it, "People regularly say to me that nobody owns the beach [...] "it's a space shared by those who turn up" (2007: 1). However, as Martin suggests above, even when small numbers of people "show up" in surfing spaces, different idea(l)s and values are made manifest. Thus just as we have seen in Chapter One that "no-one is outside or beyond geography" (Said, 1993: 7), in surfing spaces "you are never going to escape culture" (after Martin, in Capp, 2004: 221).

As outlined in Chapter Eleven, the surfing spaces script created the possibility for the establishment of a dominant culture of surfing spaces, with authentic performances realised through compliance with the script's values. Yet as Melucci suggests, in order to be successful "scripts" and "architectures" must not work in isolation, rather they are only "productive [if they are made real through] the performance of ... cultural codes" (after Horton, 2003: 64). As Horton suggests, codes are "behaviours [which] are widely and routinely recognized, if not always faithfully adhered to... and... operat[e] ordinarily as implicit and embodied principles" (ibid.). Cultural codes are therefore "rules of coherence" (after Pantzar and Shove, 2004) which "manipulate the symbolic meaning ascribed to [practices] to establish and enforce identities, lifestyles and purposes" (Yarwood and Shaw, 2010: 425, see also Mansvelt 1997; Watson and Shove, 2008). If you wish to be seen to be an authentic participant in the script of any culture, you have to comply with its constituent codes. As Thorne states in relation to surfing spaces:

> ... an[y] aspiring surfer ... ha[s] to outdo others *in kind* ...to be accepted. ... one [has] to obey the proper forms of behavior to confirm one's surfer identity
> (1976b: 271, emphasis in original).

152 *Coding Surfing Spaces*

As we have seen in Part I and Chapter Nine, there are many ways in which surfing spaces can be assembled, and many interpretations of what constitutes "the proper forms" of surf-riding behaviour. However, the meanings of the imagineers' script, combined with the power of their media architecture, transformed their own chosen behaviours from simply one way in which surfing spaces could be, into the dominant version of how surfing spaces should be. Through iterative practice (and as we shall see, the existence of threats for non-compliance[1]) this version of surfing spaces has developed from "novel", to "normal", to widely deemed as "natural" in form (after Anderson, 2021). Indeed, its script and it associated codes have become so hegemonic that, "to a large extent [they] are taken for granted by surfers" (Thorne 1976b: 272). As Pepin Silva confirms, these codes are now, "so deep in many surfers' DNA that ... they are not even conscious of them anymore" (Research Interview). A central code to this script of surf-riding is the importance of geographical *positioning* on the wave.

Positioning on the wave

Whether there are two surf-riders in a surfing space or two hundred, a basic code has emerged to determine which person (or persons) has the right to a wave. Indeed, as surfing spaces become more popular and the transience of waves remains their key defining feature, the need for (b)orders to discipline these rights is increasingly visible; as the following surf-riders explain with respect to their experiences at busy breaks:

> Mundaka [in littoral Spain] is one of the most crazy waves in the world. It's pretty hard to get it epic and the thing is that it only works on the very low tide for three hours, so as soon as it starts barrelling 150 people are out there and it is so stressed. It's pretty much like heaven and hell at the same time
> (Kepa Acero, Spanish surfer, in England and Sparky, 2014: 49).

> Twenty guys out at those breaks [like Windansea, Big Rock, Hospitals, and Horseshoe in littoral California] is fine, thirty guys is somewhat fine, forty guys and you might get a wave. Fifty guys or more, you don't get waves. People become impatient, some people might even get angry or even violent
> (Aguerre, 2015: 35).

> Topanga [in littoral California] was elbow-to-elbow and ego-to-ego, and the irony of it was that most of the people there considered surfing a religious experience and that their religious experience was being ruined by all the others surfing for the same reason. Being at Topanga on the first day of the first big winter swell was like watching pilgrims fistfight at Mecca
> (Kotler, 2006: 75).

Coding Surfing Spaces   153

As Booth, then Thorne, note above, codes exist to (b)order bodies and behaviour in surfing spaces, and a central code has developed to regulate who can access which wave (see Figure 12.1).

As Figure 12.1 suggests The Surfers' Code is broadly similar across the world (the first image is in Porthcawl, South Wales, and the second, Queensland, Australia)[2]. This code emphasises the importance of *surfer-positioning*. Surfer-positioning states that the surfer closest to where the wave is breaking occupies the crucial location; it is this location that confers the right of way to ride that wave. As *Surfer Today* explains and illustrates below (see Figure 12.2),

"In the example, surfer [on the left] has priority over the [other] surfers. If the surfer [on the left] doesn't catch it, priority switches to [the surfer in the middle]. The surfer [on the right] may only drop in if the other two don't ride that left-hand wave."

(no date)

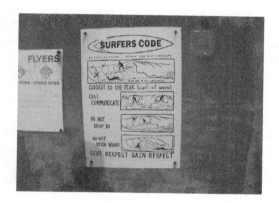

Figure 12.1a and 12.1b. The Surfers' Code (Images from author, and Lyndsey Stoodley).

Figure 12.2. Coding Surfing Positioning (Image from *Surfer Today*, no date).

In one sense, surfer positioning is the watery equivalent of joining a queue and waiting your turn as one might do on land (indeed, as *Surfer Today* states, this code imposes order "like in any civilized street system" (2021)). However, as we have seen in Chapter Three and will discuss further below, the littoral is not a terrestrial place, and the imposition of civilising (b)orders emanating from a geo-logic is in open tension with the "frontier" script which has come to define surfing spaces. Nevertheless, the existence of this code can be seen to be justified in terms of regulating the consequences of overcrowding on waves (as we have noted above)[3]. This code (b)orders access to waves based upon issues of safety, imposing clarity and accountability whilst protecting participants from risk and injury (as *Surfer Today* maintains, "When surfers don't follow th[is] basic commandment… accidents, injuries, and misunderstandings happen" (2021). Any individual who wishes to be considered a "good citizen" of surfing spaces would therefore assume this code articulates a common-sense view and are likely to align their behaviours in line with this orthodoxy. In short, they would comply, choosing to "give respect [and] gain respect" (see Figure 12.1), and through so doing be recognised as legitimate member of this space.

Yet it is possible to suggest that this (b)order imposes more than simply safety on the waves. As we will see, positioning on the lip is directly connected to issues of local knowledge, strength and ability, craft-choice, and attitudes towards wave-sharing. In sum, the imposition of this code is not simply a form of common sense, it is a direct consequence of and contributor to particular ideologies which define authority, privilege, and surfing style. As noted above, surfing spaces are not terrestrial in nature, and thus surfer-positioning is more complex than simply being the watery equivalent of joining a "civilized street system". Waves are not like bus stops or traffic lights, they do not always exist in precisely the same location each time (see Chapter Three). As a consequence, it is those surf-riders with local knowledge of that surfing space which have greater opportunity to sense where the most likely take-off point is going to occur. In a similar way, choice of craft, strength, and expertise will also influence the ability of any surf-rider to occupy the location of the lip.

Stand up paddle boarders, for example, may be able to see the signs and shadows of forthcoming waves better than other surf-riders due to their elevated position, similarly SUP-ers' or surf-kayakers' ability to use a paddle may help their capacity to catch a break at minimum notice. If surf-riders are using similar craft at one surfing space (for example short-boards), then those individuals who have greater physical strength and skill are likely to have the advantage to execute a late drop and occupy prime position. In these ways, surf-rider positioning is a code that at once imposes a greater degree of safety in surfing spaces, but also establishes a hierarchy of surf-riders. As a consequence of this code, experienced, strong, local, surfers, using particular crafts best suited to that break occupy a privileged position. Over time, as these riders are more likely to anticipate the location of the lip, and therefore more likely to catch waves, their privileged position is consolidated; the "wrong type" of surf-riders catch fewer waves, they are consequently less able to improve their expertise, and are ultimately marginalised from the surfing space.

Squatting rights and surfing solitaire

By identifying the prime location for catching a wave, the code of surf-rider positioning suggests that although a range of take-off points exist on any wave, they should only be used if prime position is vacated (see Figure 12.2 above). This code therefore also prescribes a certain philosophy of surf-riding which may not be obvious at first glance: the ideal of "one surf-rider, one wave". As Greg, a surfer from South Wales puts it:

> when you surf, and you caught the wave, right? You know that wave is yours now, it's not for sharing, it's just your wave... So it's like, that's the rules, you know, don't drop in. If you're on a wave already, there's not enough room for two people. There's only enough room on one wave for one surfer at the end of the day
>
> (Research Interview).

As Greg confirms above, a further implication of the surf-rider positioning code is that waves are not for sharing. As *Surfer Today* sums up, "As a general rule of thumb, you'd say that surfing is 'one man [sic], one wave'" (no date). In this interpretation we can identify the influence of the surfing spaces script. As Severson put it, "In this crowded world, the surfer can still ... *be alone* with the surf" (2014: no page, emphasis added; see also Chapter Ten). This idea(l) is secured in practice if and when a surf-rider is able to occupy the lip; if they can do so then that wave becomes theirs, their positioning bestows the opportunity to conclude their quest, obtain their grail, and ride a wave alone. This codification therefore directly defends the surfer's interest; it protects their opportunity for escape and their opportunity for stoke. As such this code valorises the individual on the wave (it realises and reinforces the view expressed by Winton in Chapter

One that "eventually, there's just you and [the wave]" (2008: 85), and nobody should get in the way of this individualised encounter.

This take on surfing spaces suggests that being located on the lip grants a form of territorialised ownership of the wave. As Ford and Brown (2006: 135) put it, that wave becomes that surf-rider's "possessional territory" (after Goffman, 1971) and respect should be shown by others to this ownership-through-occupation. In a similar way to the transference of highway controls from the terrestrial to the littoral, this code imposes a proprietorial geo-logic to surfing spaces, suggesting that,

> the space around the surfing body and specifically the line they are taking on the wave become 'theirs' [i.e. that individual surfer's] for the short period of the ride on that part of the wave
>
> (Ford and Brown, 2006: 135).

At once this proprietorial approach secures a number of outcomes. As a space now understood to be secured for an individual rider, it becomes their place for action. It is accepted that as they have secured the "best"-positioning on the wave they are absolved from responsibility for the well-being of any other surf-rider who happens to be in a sub-optimal position (it is assumed that the latter should know the code and have to take the consequences of their non-compliance, see below). The best-positioned surf-rider is thus liberated to execute their moves and attain a "flow", "converged", or transcendental state[4], but they must also demonstrate their competency as they do so to their watching audience. In this way the wave as a place for action also becomes a stage of cultural performance; "appropriate" forms of knowledge, reaction, and riding are judged by others in the water and on the shore, and in this way the code of surf-rider positioning is reinforced through cultural spectacle.

If any surf-rider fails to acknowledge the code, there are consequences. As Horton suggests (above), although cultural codes are widely and routinely recognized, it is always possible that they are "not always faithfully adhered to". If a surf-rider should "jump the queue" and drop in on a wave that is not "theirs' to ride, it is called "snaking". Scheibel suggests that this terms explicitly associates this code-violation with Biblical treachery, with the perpetrator or "serpent being symbolically linked to the Garden of Eden... to have been 'snaked' is to have been sinned against" (1995: 253, cited in Ford and Brown, 1996: 79). The quasi-religious tone which defines this violation of the (Edenic) wave emphases the strength of feeling against such deviancy. If the code of surf-rider positioning is "disrespected" in this way (following Figure 12.1) then those who break it (whether intentionally or otherwise), "should expect consequences" (*Surfer Today*, no date). In line with the "rugged individualism" of the surfing spaces script (see Chapter Ten), this violation is likely to, "end up in unpleasant discussions, insults, and fights" (*Surfer Today*, 2021). "Insults" may involve the "snake" being ridiculed as a "kook", or a "barney"[5], and coerced to future compliance through threat (and sometimes the realisation) of violence. This violence may be to property, or may be personal. For example, a kook's car may be vandalised by enforcers of the code, as Wade articulates in relation to Cornish surfing spaces:

> The [threat] is made with some humour by 19-year-old St Agnes local Josh Ward. As with many Badlands' surfers, Ward comes from a family that has surfed in the area for generations....As the bleach-blond-haired Ward says: "My whole family surfs, from my nan to my little brother". But then comes the Badlands punchline: "But watch out, my nan shreds low-tide Aggie and she'll wax your windscreen if you drop in on her...."
>
> (Wade, 2007: 86).

We will see in Chapter Thirteen how violence to people has become a crucial element in ensuring compliance with surfing space codes but with respect to "snaking" in particular, the following reflection by Duggan outlines how in Hawai'i, the importance of showing respect to the code is not limited to abiding by it, but is extended to enforcing it. As this example suggests, if a surf-rider fails to enforce the code, they may appear to be as treacherous as those that transgress it:

> Keane noted the etiquette on the water [at Pipeline, littoral Hawai'i]. He saw a body boarder dropping in on a surfer, forcing him to abandon his wave. The surfer gestured at the culprit but that was it. What got Keane, though, was that when the surfer eventually came in, his friends were waiting to berate him. They wanted to know why he hadn't beaten the shit out of the bodyboarder
>
> (2012: 115).

Code transgression

As this chapter has demonstrated, the code of surf-rider positioning enforces the script of "one rider, one wave", putting into practice the assertion that "more than one surfer on a wave greatly diminishes the enjoyment of both" (Ponting and O'Brien, 2014: 100). Yet as we have seen in Chapter Nine, a variety of endogenous surfing spaces celebrate wave-sharing through their histories, and although this pursuit may privilege particular craft- and wave-types, there is no "natural" or "normal" common sense which means that sharing waves inevitably "diminishes the enjoyment" that can be obtained through engaging with them. Here we can suggest that the surfing space script (and its associated power relations) has a direct influence. Rather than promoting individualism and self-determination over co-operation and conviviality, the code of surf-rider positionality could, for example, bestow upon the rider positioned at the lip the right to decide whether they share the wave or not. At face value, such consensual wave-sharing would align more closely with the "freedom" suggested by the overall script, yet it would also licence the transgression of its associated positioning code. As such, this interpretation would reveal the code to be far from a natural or common-sense inevitability, but rather a chosen (b) ordering mechanism that is integral to a script which accords advantage to a few. Reclaiming choice therefore creates risk and uncertainty, not simply to the practice of surf-riding on some waves, but to the hegemonic position of the surfing spaces script as a whole.

In recent years, transgression of the code of surf-rider positionality through consensual wave-sharing has become increasing apparent. Such wave sharing remains unusual, but with the rise of social media, on those occasions when it does occur, it gains wide interest. Wave sharing may occur in littoral locations where the wave is suited to mass "drop-ins"[6], where the assembled surfing space is open to conviviality (see Greenaway, 2020), or when individual surf-riders are willing to "ride with it" when unintentional transgression occurs (see for example *Surfer Today*, 2014). On the majority of occasions, however, if consensual wave-sharing occurs then those involved are often ridiculed (perhaps as "kooks on foamies", see Beefs TV, 2019), or accepted as novel(ty) exceptions which do little but prove the coded rule (*SoCal Surfer*, 2019).

Conclusion

As we have seen in Part II, surfing spaces have been scripted as an open frontier. Despite this, in practice surfing spaces have been (b)ordered through codes of surfer-positioning. This code imposes clarity, accountability, and safety onto surfing spaces and has been naturalised on a global scale. Although this code offers the opportunity for any surf-rider to obtain optimal positioning on a wave, it is more likely that seasoned locals have greater ability to anticipate where a wave may break. Thus whilst this code imposes safety on a wave, as it does so it creates a surf-rider hierarchy which marginalises the novice surf-rider and establishes surfing spaces as the domain of the local individual. However, as we will see, this local entitlement sometimes depends less on the *positioning* of that local surfer, but more on their *positionality*. Chapter Thirteen goes on to explore the importance of positionality; it does so in relation to the code of surf-rider provenance, examining how this code can consolidate inequality in the line-up, and the strategies used to impose and perpetuate this (b)ordering of surfing spaces.

Notes

1 As *Surfer Today* notes, "Surfers who break [the codes] should expect consequences. Breaches of… rules usually end up in unpleasant discussions, insults, and fights" (no date).
2 Indeed, according to *Surfer Today* (no date), this Surfers' Code "applies to all line-ups of the world." The hegemony of this American script is so paramount that even scholars like Comer can take it for granted; as she does when she labels the Surfers' Code "a bill of rights" for "global surf citizenship" (2010: 25), implicitly accepting the apparent inevitability that a US approach to code-making is ideal in form and definition to be applied everywhere.

3 Due to the rising prevalence of "overcrowding" in some surfing spaces some writers compare the codes which (b)order surfing spaces to those regulating the use of a traditional "commons resource". As Barilotti states: "The outstripping of a surf spot's carrying capacity closely follows biologist Garrett Hardin's classic 1968 essay "The Tragedy of the Commons" in which he illustrates how unchecked populations and capitalism will eventually graze a shared resource – whether it be grass or waves – beyond sustainability" (no date).

Here surfing spaces are compared to a resource that is, in the popular imagination at least, open to all potential users and, as a consequence, risks overexploitation in the long term (as Hardin famously argued in 1968). As a "free zone", there may be some merit in the comparison to the "tragedy of the commons". The (b)ordering of surfing spaces can be understood as an attempt to regulate access to a straightforward geographical territory and protect it from overuse. However, surfing spaces are not commons in the same way as terrestrial reserves may be. As Nazer (2004) identifies: "Surfers' commons are different from a fishery or a communal pasture. In fact, it is not even immediately obvious if there is such a thing as "over-surfing". Over-fishing today can lead to no fish tomorrow. In contrast, today's surfers cannot change tomorrow's waves. There may be good surf tomorrow no matter how crowded the waves were last week" (2004: 658, 659).

Thus whilst some terrestrial resources have a degree of material permanence (e.g. silver or gold), many can be used up (e.g. coal through burning, or pasture if overgrazed). The surf zone and singular waves, in particular, are different. They have little material permanence, being both capricious and unpredictable. Although they cannot be "used up" in a material sense, they can suffer from cultural overuse when more than one surfer attempts a ride (as is discussed later in Chapter Twelve). Although it may be tempting to compare surfing spaces to a renewable resource (similar perhaps to wind, tidal or geothermal energies), surf resources cannot be stockpiled or transformed into "hard" energy; they remain ephemeral and transient. Surfing spaces – both materially and affectively – are thus a resource "of the moment" and for use by the individual "in the now".

4 This code reminds us, therefore, that surfing spaces are (b)ordered in part not simply to claim ownership of a wave in a terrestrialised and territorial sense, but also to regulate access to the relational sensibility it affords them (see Anderson, 2013).

5 As Duane outlines, a kook is a, "a universal surfing term for the unknown, unimpressive other, and suggest[s] the ridiculous jerking motions of an incompetent surfer; Surfer magazine even runs a regular cartoon about a hapless idiot named Wilbur Kookmeyer. Barney, meaning roughly the same thing, seems to derive from Fred Flintstone's little buddy, Barney Rubble" (1996: 12). As Heller goes on, "Kook means 'beginner surfer' [but] it is not a neutral term; it carries a slug of derision, a brand for the clueless, for those without hope, without grace, without rhythm. To be a kook is to be consigned to a kind of beginner's hell" (2010: 2), and as Beaumont sums up, the term kook, "originates from the Hawai'ian term Kukai which literally means 'shit'"" (2011: 156).

6 As Bromley describes, "waves like Waimea, Sunset Reef in Cape Town, and the shoulder of Mavs are pretty much open game for drop-ins because they don't have much of a wall. The general angle of people dropping in is pretty straight. But then you get to Jaws, Dungeons, or the outer bowl at Mavs, and it's totally a one-person wave. On these waves, you're engaging the rail as early as possible and pulling up onto the face or into the barrel to make the wave. If someone were dropping in down the line, the inside surfer would be t-boning them in a very critical situation" (in Rode, 2020: no page).

13 Codes of Surf-rider Provenance

Introduction

As we have seen in Part II, a surfing spaces script has been imagineered which translates the stoke of surf-riding into a narrative of freedom and the open frontier. Yet as Hough-Snee and Eastman point out, "surfing… is a profoundly complex… practice, rife with contradictions" (2017: 2). So, whilst the script of surfing spaces sells the ideology of freedom for all, Chapter Thirteen has outlined how the code of surf-rider positioning (b)orders these spaces to privilege some littoral-humans, including those with local surf-riding identities. In this chapter we examine how this code of surf-rider *positioning* is consolidated and perpetuated through the first code in a set of surf-rider *positionality*.

We have seen in Chapter Eight how surf-riding is "eventful" for many of those who undertake it, establishing and consolidating a geo-cyborg identity which co-constitutes the surf-rider with littoral locations in general, and some spatial assemblages in particular (Chapter Five). The code of surf-rider provenance refers to the (b)orders of privileged access that are claimed to these specific spatial assemblages, based on the experience, competency, and co-constituent relations that local surf-riders have in and with their "local" breaks. This chapter suggests that this it is common for this sense of local co-ingredience to be accompanied with a concern that incoming surf-riders may disrupt their geo-cyborg relations. As a consequence, outsiders are often treated as "strangers" by locals, with protectionist tactics used to deter and (b)order their presence. However, we will also see how the code of surf-rider "provenance" relates not simply to a place of origin or habitation, but also to "authenticity" and "quality"; we will see how it is possible for those surf-riders who do not enjoy local status can nevertheless prove a degree of provenance to local surf-riders in other "respects" – through performing an awareness of the codes of positioning and positionality, and through demonstrating a concomitant ability on the waves.

Provenance

A key (b)order which seeks to regulate the open frontier of surfing spaces is the code of surf-rider provenance. As we will see, this code has many facets, but the first and most obvious refers to the privileged access to waves which is assumed by those who live close to and regularly surf them. Reflecting the process of person-place constitution (see Chapter Four), through consistent and recurrent engagement with a local littoral, surf-riders come to gain knowledge, expertise, and understanding of how this surfing space works (both geographically and culturally); as a result, the local littoral comes to be constitutively co-ingredient with their sense of self. Although a sense of connection can be easily created (see Chapter Eight), such local co-constitution is hard-earned; to be at one with a littoral location requires an on-going commitment to be made and performed. Thus to be a local (according to Finnegan at least) one must carry out,

> the close, painstaking study of a tiny patch of coast, every eddy and angle, even down to individual rocks, and in every combination of tide and wind and swell – a longitudinal study, through season after season[. This] is the basic occupation of surfers at their local break. ...Getting a spot wired – truly understanding – can take years
>
> (2015: 75).

Wiring oneself into a local surfing space confers practical benefit to the individual earning this status. As we have seen in Chapter Twelve, due to this hard-won co-constitution locals are more likely to have the vernacular knowledge which enables better positioning when the waves come. Yet beyond this, there is nothing essential or natural as to why locals should gain privileged access to the waves near their home, so why in practice does the code of provenance exist? Why are locals deferred to by other surf-riders when the waves arrive? Why do people without local status choose not to openly compete for waves if local surf-riders attempt to catch them, or pull out (even if they are better positioned) in order to respect the provenance of the local rider at their "home" break? In a script which suggests that littoral locations are open frontiers and celebrates the rugged individualist who single-mindedly searches for stoke, why has the code of local provenance become naturalised at many littoral locations around the globe?[1] This chapter progresses by exploring the reasons for the normalisation of this cultural code.

Normalising the code of surf-rider provenance

In many places the code of surf-rider provenance exists as many surf-riders respect the knowledge and ongoing fidelity of locals to their specific littoral assemblage, and translate this into privileged access to their surfing space. The code of surf-rider provenance is thus an (in)formal acknowledgment of a surf-rider's geo-cyborg identity, oriented through practice and ongoing performance at a specific littoral location. The code recognises the co-constitutive nature of person and place, and the ways in which through incoming and outgoing over the seasons, a surf-rider's body is extended and dispersed to become part and parcel of the local surfing space assemblage. Yet in these cases, the establishment of this code did not happen naturally, indeed, the code emerged through ongoing and concerted conflict between existing surf-riders and newcomers, all influenced by the imagineers of the surfing spaces script.

As we know, surf-riding has grown in popularity since the 1950s, linked directly to the success of the imagineering process. As the imagineers sought to enlarge the constituency of surf-riding in order to gain profit, they popularised the practice and – to some extent at least – enabled newcomers to "drop in" (albeit metaphorically, see Chapter Twelve) to local littorals. Through providing authoritative guides which detailed the types of waves prevailing at any given location, and advising on the type of boards and riding styles best suited to ride them (see for example, Alderson 2008; Nelson and Taylor, 2008), the media architecture of surfing indirectly under-cut the hard-won knowledge of the "local" surf-rider. The imagineers enabled newcomers and outsiders to gain access to waves without earning knowledge through practical experience.

However, as we know from Chapter Eleven, the original imagineers of surfing spaces were also surf-riders themselves (and as the surf media has developed, this model of the surf-lifestyle entrepreneur has tended to dominate the industry, see Stranger, 2011). As such, these imagineers appreciated from their own experience the geo-cyborg relations that constitute the local surf-rider, and felt strongly the consequences of the growing popularity of surf-riding that they had imagineered into being (see Chapter Eleven). Thus they were faced with a dilemma generated by their combined status of surfer and capitalist: to reduce the number of surf-riders would mean to undermine the growth of their business, but to side against local surf-riders would mean to surrender their own local privileges as surf-riders[2], and alienate their core market as entrepreneurs. The way out of this dilemma was to perpetuate a particularly reactive and "mixophobic" stance to non-local surf-riders.

The threat of the "stranger": "If you don't live here, don't surf here" (Lewis 2009)

> Legend says that one summer day in the early 1965, a group of "locals" who were expecting good waves attempted to block the canyon roads in order to keep surfers from the Valley – the Vals – off the beach. Over the next thirty years, "Vals" were routinely victimized by "local" surfers at numerous Southern California beaches
>
> (Scheibel, 1995: 256).

The case of the "Vals", as Scheibel notes, has become "legendary" within the surfing space script as an example of local surf-riders wishing to keep non-locals from their waves. What is key to this example is the way in which the dominant surf media and imagineers responded to this incident. This material (b)ordering process (blocking roads) and the subsequent physical and verbal harassment (routine victimization) was not condemned by the media architecture; although it was not overtly condoned, many efforts were (and are) made to explain and justify these actions. As Drew Kampion who had direct experience of the ongoing Vals situation, sums up:

> I was living over the hills in the San Fernando valleys, and I was one of the people... identified as being one of the kooks from the "mainland", from "over there", who was ruining [the locals'] world. And so I kind of ... could definitely empathise with [the sense of threat felt by local surf-riders]. All of a sudden this wave [you] could ride [alone] became literally populated by hundreds of people. [It became] pretty trashy, pretty destructive. A paradise was taken away from you, so [we tended] to feel very empathic with [the locals' view]
>
> (Research Interview).

Such empathy with the attitude and tactics of local surf-riders when faced with non-locals can be seen to effectively legitimise and validate them. Indeed, "Vals" has now found widespread use in US surfing spaces as a derogatory term denoting "outsiders" who have an inferior status to locals (and as a term is not dissimilar to the UK label "grockle" which denotes individuals somehow less deserving and less "in place" in a local's territory (see Cresswell, 1996)), and local surf-riders often display suspicion to non-locals attempting to surf "their" waves.

> "Wait a sec", came a soft voice from the back of the room. "I live in Northern California and I drive a lot looking for ridable waves. I pay attention to the wind; I keep an eye on shifting sandbars; I listen to the weather radio all the time." He paused for a second and chose his words carefully. "When I paddle out and see a bunch of guys that I don't recognize at a spot that's taken me years to figure out, I feel... almost violated. Especially if they're not respectful
>
> (Sanders, no date).

As Sanders suggests above, many locals treat newcomers to a littoral location as "strangers". At one level, this label may be simply employed as a colloquial term which refers to individuals who locals do not yet know. However, it is also possible to read this term as part of a process of "Othering" (following Said, 1979) that is implicitly legitimized and perpetuated within surfing spaces.

"Strangers" (following Simmel, 1950), can be understood as individuals who are considered to be "out of place" (after Cresswell, 1996) in a given location. As Bancroft explains, the "Stranger" is considered to, "disturb... [and] threaten to overthrow that psychological unity" which holds a person and place together (1999: page). The "stranger" is thus at once a cultural and geographical being, an individual who is not from "here" and as such automatically assumed to be illegitimate to be "here". Provenance is primary in this perspective, it confers a social identity (i.e. who you are is where you're from) and with it a cultural entitlement, in this case to littoral access ("you" will never be as deserving as "us" to ride these waves). From this attitude to the unknown and threatening other comes a prejudice that "governs" (after Watson, 1983: 388) surfing spaces. This prejudice is named by sociologist Zymunt Bauman as "mixophobia" (1995: 221), the process where individuals are not treated equally as fear and a loyalty to the "in group" results in "intolerance, and sometimes much worse" (after Barnes, 2005: 72) towards the "out" group. In the case of the code of surf-rider provenance, this mixophobia is bluntly summed up by the following phrase: "If you don't live here, don't surf here" (Lewis 2009).

> locals feel at home at the break and feel invaded by unfamiliar faces, they sometimes react in a negative manner to newcomers
>
> (eHow Sports and Fitness Editor, no date).

As with other aspects of the surfing spaces script, the code of surf-rider provenance has spread to many surfing spaces around the globe[3]. Yet although locals may wish to extend their sense of territorialisation of the land to the surf zone, in many cases it is difficult to use conventional (b)ordering mechanisms (such as blocking roads) in the liquid and fluid materiality of the sea. Surfing spaces cannot be straightforwardly gated (although in some cases the beach adjoining can become privatised), so how do local surfers (b)order this code in practice?

> Surfers have a long tradition of spray-painting 'Locals Only' on footpaths and rock walls to let people know that particular beaches and pieces of turf are 'theirs'
>
> (Evers, 2007: 2).

> Oliver Poole, a reporter for the UK Daily Telegraph, took a long-awaited trip down the California coast recently. He paddled out at Steamer Lane expecting to commune with a mellow tribe of California sea nomads, but instead had his cheery British "'ellos" returned with stinkeye and hostility. Poole promptly wrote back home of his disillusionment; "It was a quick lesson in the modern rules of Californian surfing: keep off our beach, don't mess with our waves..."
>
> (Barilotti, no date).

The linguistics of "vals" or "grockles" are supplemented in practice by a range of other (b)ordering measures. From behaviour including the "stinkeye" and "hostility" alluding to by Barilotti (above), the code of surf rider-provenance is enforced through verbal and physical intimidation – and sometimes violence to both people and property.

> They might wear relaxed-looking clothes and carry around surfboards, but some surfers have the same mentality as football hooligans
>
> ("Tony", surfer, cited in Rich, 2006: 1).

Localism's effects vary... from a dirty look and a waxed windscreen to criminal assault and expensive car vandalism

(Anderson, 2007: 54).

As the above quotations outline, local surfers may use practices on land to restrict access to "their" surfing spaces. Through labelling and signing the beach as "locals only", or by informing incomers that they are unwelcome (through defacing windscreens with surfboard wax or vandalising cars) locals attempt to dissuade "others" from accessing the beach. In some cases, however, access itself is not (b)ordered, rather the surfing space itself becomes the place where incomers are dealt with. As the following interview respondents from South Wales recount:

Q: Is Localism ever an issue?
A: Yeah, _____ [surf spot in South Wales] can be a bit hostile if the waves are quite good and you normally get the locals out in a big group and it is quite hard then to catch the waves for yourself
Q: What happens?
A: If you take off on a wave, they will just take off on a wave behind you and they will make sure that you are the one that falls over

(Anonymised surf-rider, Research Interview).

I've surfed in a lot of different places. I've just come back from Morocco. I do tours of France and Spain on the way back. But parts of France are horrendous. I mean, you get people on the beach fighting continuously

(Anonymised surf-rider, Research Interview).

Localism... does exist, especially round here. If you paddle out and try to dominate, the local guys will have words with you....The guys will just close ranks. If they don't like you they'll bunch up and close the peak off. You can paddle as hard as you like but you won't get a wave

(Lascelles, referring to Cornish beaches, cited in Wade, 2007: 88).

A range of tactics are therefore employed to deter non-local surf-riders from destabilising the privileges of access accorded by the code of surf-rider provenance. Tactics such as physical threat and violence could, of course, be classed as intimidation and assault. However, it is rare for formal authorities such as the police to become involved when locals discipline their surf breaks. As mentioned above, in line with the broadly "empathic" view towards mixophobia, within the media architecture of surfing spaces physical violence and criminal damage are rarely condemned. Indeed, in line with the script of rugged masculinity, such tactics are couched as part of an honourable crusade, upholding their code against imposing others, as Barilotti puts it,

166   *Codes of Surf-rider Provenance*

> There is such infinite dirty pleasure in burning a righteous [outsider]. [It can] take on the grandiloquent nature of a storied duel. Your cheek is stinging and one's honor is screaming for satisfaction
>
> (no date).

In these ways, the range of tactics employed to (b)order the code of surf-rider provenance can each be seen to an act of power. As each takes and makes place, it does so in both a regressive *and* a progressive way, depending on your positionality in relation to it. In one sense it defends idea(l)s and visions of the local geo-cyborg, but on the other it is prejudicial, mixophobic, and sometimes violent to those it determines to be "strangers". So for those exercising such tactics, outbreaks of violence are explained in relation to the absence of honour and respect to the traditions of that place which "strangers" bring about, as Jack Kingsley a Bra Boy (from Maroubra, Australia, a bay infamous for uncompromising (b)ordering) puts it:

> You've spent so much time in a one place, day in and day out. You've lived your life there, and it's a respect thing, just from one person to another.... It's an unspoken rule... you just expect people to show some courtesy and behave when they come to your home
>
> (Jack Kingsley Bra Boy, in Surfline no date).

Thus even though these practices directly undermine the symbolic freedoms of the surfing spaces script, they nevertheless become cultivated and mythologised as part of it. Motivated by a lust to defend their frontier and a desire to protect their stoke, the willingness to ruggedly transgress "normal" (b)orders of civilisation become aligned with the broader surf-rider archetype. As this code and its tactics become known even the suggestion of violence in a particular surf spot (rather than the practical likelihood of it ever occurring) is enough to (b)order the surfing space; stories of transgression and punishment trade on the image of groups like the Bra Boys and Da Hui (on the North Shore of Hawai'i, see also Chapter Sixteen), and are enough to keep outsiders away, as the following examples go some way to illustrate:

> "We don't really beat people up round here. It's a bit of a myth. Besides, a lot of the blokes aren't tough enough to back up what they say. It's a myth. But it's a myth that's worked. Chapel Porth and Trevaunance Cove are two of the hardest places to get a wave in Britain". There was an unmistakable twinkle in Lascelles' eyes as he told me this
>
> (Lascelles, in Wade, 2007: 88).

To paddle out at, say, Chapel Porth, Trevaunance Cove or Porthtowan is, if you believe the hype, not far short of an act of hubris. Chris Nelson sums up the feelings of foreboding in Surfing Britain: "The world has many Badlands. To me, as a young grommet surfing Yorkshire in the late

1980s, there was only one true Badlands, the 'locals only' breaks around St Agnes. The name alone was enough to keep us away when we made the occasional foray to the south west. A magazine article there, an urban myth here. Tales of spontaneous acts of violence, regular drop-ins [see below] and a very warm welcome but not of the pleasant variety. We didn't think our scruffy blue Datsun would really benefit from flat tyres and a 'locals only' wax job on the windscreen. Here, there were other beaches, other waves. We never ventured down to check the place out"

(Wade, 2007: 85).

At once therefore, the code of surf-rider provenance exists as an (in)formal acknowledgement of the hard-earned co-constitution of local surfers in particular locations. But this acknowledgement has not arisen through a natural process of mutual and reciprocal respect between surf-riders, rather it has been imposed through a selfish and mixophobic attitude demonstrated by locals, and tacitly (and sometimes explicitly) legitimated by the broader architecture of the surf media. As such, the code of surf-rider provenance not only draws attention to the pride that local surf-riders feel towards their surfing space, it also reminds us of the protectionism applied by them towards it, a protectionism that ensures that the code of surf-rider provenance is maintained in practice.

Provenance plus: A code of authenticity and quality

However, the prevalence of the code of surf-rider provenance does not mean that if you are a non-local surf-rider, you cannot catch waves at any surfing space beyond your local. Access can also depend on a number of other factors, including the nature of the specific surfing space assemblage, it's carrying capacity, the normal number of surfers, their craft, ability, as well as local attitudes. Reflecting the definition of "provenance" itself, a working understanding of these variables will help non-local surf-riders align with the principles of this code. If surf-riders can demonstrate their respect for these variables, alongside the code of positioning (see Chapter Twelve) then they can be accorded a degree of provincial "authenticity" and "quality" on local waves.

Definitionally, the term provenance not only alludes to the origin of, in this case, a surf-rider, it can also refer to attributes of "quality" and "authenticity". These traits can be accorded to a non-local surf-rider by locals in any given littoral if they demonstrate respect to the codes of positioning and positionality as outlined thus far. Thus, if a non-local surfer abides by the code of surf-rider positioning (i.e. they don't drop in), and the code of surf-rider provenance (i.e. they display respect to local surf-riders' privileged status on the waves), it is possible to gain a degree of acceptance within a local surfing space. As one Welsh surfer outlines:

> We tell the young surfers look, if you go away surfing you gotta play the game, you just can't go wading in, thinking you're a pro cos it won't work

and... you'll have bad vibes, let's say. Basically you go in ...just don't move, just watch [local surf-riders] get some waves, watch them get a couple, and when you've seen one of them and he's had a really good wave and they're really happy, that's probably a good time to say hello to him, you know "how's it going?", "yeah nice wave" you know. "What board you riding?", start a conversation, get your foot in the door. You can go be the outsider who says nothing to none of them, or you can go and make friends. And it's a skill, there's an art to that in itself....If you show disrespect, you know you can be a long way from home, on your own and have a really bad time

(Anonymised Research Interview).

Performing respect to local surf-riders can therefore confirm a non-local's authenticity and quality as a surf-rider: their actions demonstrate they understand what being a local surf-rider feels like, what efforts their geo-cyborg relations took to generate, the lengths they may go to protect them, and what respect this should confer upon them as a result. As such, through performing this respect in practice, they are one step closer to being tolerated within a local's surfing space. However, quality and authenticity are also attributed in relation to the act of surf-riding itself; if non-locals are able to catch waves in new spaces, it is vital that the waves that are bestowed upon them are not "wasted". Non-locals will have to ride these waves well to take the next step of being deemed a surfer with provenance; as the following Cornish surf-rider states:

> I think your sessions [in non-local surfing spaces are] made easier if on your first wave you get a good wave cause everyone then knows that you can surf and then you'll be treated with a level of respect and dignity in the water. If you go out and you mess up your first wave then everyone assumes that you can't surf and then they'll drop in on you more. If you get a good first wave and you feel like "yeah I'm good, I'm alright, the guys know that I can surf"

(cited in Beaumont, 2011: 179).

As waves are routes to relational sensibilities, a wave that is not surfed well is a wave wasted, thus on busy breaks, locals may not take kindly to a non-local failing to ride a wave in line with its capacity. Indeed, wasting a wave may also incur a risk to the wiped-out surf-rider and others in the space, as American surf-rider Jacob Malthouse states,

> Often a local won't explain the fact that they're worried about you hurting yourself or someone else to you [because they know the waves are simply beyond your ability], they'll just tell you to get the F**k back to the beach, leaving said kook feeling miserable and angry

(1999: no page).

The code of surf-rider provenance thus implies that surf-riders must have a good working knowledge of their own surfing ability, and its alignment with the waves that are being paddled for. Ultimately, these skills cannot be undercut by easy-to-gain data from the surf media, they can only be learned through practice. Thus for any surf-rider in any littoral location, the code of surf-rider provenance suggests that it is best to initially,

> go out… on small days so you can learn the currents, rocks and shape [of the surfing space assemblage. On these days] a lot of the time the locals won't go out because it's too small, or they'll be in friendly mood because small waves also mean small crowds
>
> (ibid.).

By accreting local knowledge in this way, social capital can be generated between locals and non-local (or new) surf-riders, and all aspects of the code of surf-rider provenance can be respected and earned through practice.

Conclusion

As we have seen in the last two chapters, the illusion of surfing spaces imagineered by the dominant script is far from the practical reality. The code of surf-rider positioning confers privilege to local surf-riders, and this is consolidated by the first code of surf-rider positionality: the code of local provenance. This code acknowledges and rewards the time and commitment surf-riders give to a specific littoral, and recognises the eventful and geo-cyborg identity such practical commitment generates. This recognition is ensured by the protectionist tactics used by locals to defend their space from "strangers". This protectionism panders to the "myth" of rugged individualism in the surfing spaces script and naturalises the prejudicial politics of mixophobia. Provenance thus becomes a key "marker of difference" (after Anderson, 1999: 4) between "authentic" and "other" surf-riders, imposing and justifying unequal power relations between locals and non-locals; as Mark "Scratch" Cameron, a Scottish surf-rider puts it in respect to non-locals as his home break:

> If you sit and wait your turn [here] I'm happy to give you waves. I'll maybe get most of the set waves, that's how it works, but if you're nice and friendly, come on in
>
> (Currie, 2021: 92).

The code of surf-rider provenance has thus become a "taken-for-granted reality" (after Kobayashi, 2013: 57) in surfing spaces, as Cameron notes above, it is now just how they "work". The normalisation of this code nevertheless ignores other ways in which surfing spaces could be (b)ordered, scripted, and supported by its broader media architecture. We will see in Chapter Fourteen how alternative ideas of craft could lead to different ideas of space and

170 Codes of Surf-rider Provenance

territorialisation, and in Chapter Fifteen how different attitudes to the "other" may promote conviviality and community over conflict. However, the influence of the dominant script nevertheless remains, and in the next chapter will we examine how this creates further inequalities through the code of craft.

Notes

1 It is worth noting, however, that this code is not universal and is not applied everywhere, as we will see in Chapter Sixteen.
2 The nature of the surfing spaces script legitimizes a selfish attitude to the search for stoke, as surf-rider and -writer Tom Anderson admits, although "everyone has equal rights to surf wherever they want, ... we're a selfish tribe, and hate the prospect of wave sharing" (2007: 54).
3 As one anonymised surf-rider puts it in interview:
"I've been all over the world surfing. From Hawai'i to Fiji to Australia, the usual, like 7 or 8 times, and when you go to these new places... you just can't rock up and like think you can go in the water and take all the best waves before all the locals. They don't like it, it doesn't work like that, that's not how it works at all" (Research Interview).

14 Codes of Craft

Introduction

As we have seen in Part II, codes of surf-rider positioning and surf-rider positionality (b)order surfing spaces. These codes at once complete the imagineering process, implementing the dominant script of surfing spaces in practice, but they also exist in tension with it, perpetuating a politics of surf-riding which privileges some types of surf-riders and marginalises "others". We have seen how these codes (b)order surfing spaces in terms of positioning on a wave (Chapter Twelve) and in terms of local provenance (Chapter Thirteen); in this chapter we turn to the second code in the set of surf rider positionality: the code of (surf-rider) craft.

As we have seen in Chapter Five, there are a range of technologies available to ride the breaking wave. Each can be thought of as a prosthesis, extending and enabling the human to come together with surfing spaces and have different experiences and encounters. The range of surf-riding technologies (from body-boards, to kayaks, to short- and long-boards) enable different waves to be surfed and different ways of surf-riding to be generated; as a whole they broaden and diversify the "craft" through which surfing spaces can come into being.

In this chapter we examine how the prominence of particular craft materialities (i.e. the short-board) and craft styles (i.e. speed) have come to further (b)order how dominant surfing spaces are understood in practice. We will see in turn how these crafts have come to privilege a particular age group and produce a particular culture of difference and conflict. More specifically, this chapter argues that these crafts have instilled a politics of verticality into surfing spaces, involving an emphasis on, firstly, being vertical to ride, and secondly, riding vertically (i.e. not just along, but also up and down a wave). As with other (b)orders emanating from and re-producing the surfing spaces script, this code of craft has created tensions, and alternative voices are increasingly being heard which challenge the dominance of this code and ultimately call into question the viability of the surfing spaces script in the future.

DOI: 10.4324/9781315725673-14

Craft

In this chapter, the idea of craft can be considered as both a noun and a verb. As a noun, a craft is a material thing, an item we can touch and engage with, it is a unit that is palpably sensed between our fingers, or through our feet. Yet in light of the relational approach of this book, this materiality is not always straightforward and unitary, it can be assembled through its engagement with other "nouns" (i.e. other things and materialities) in a set of ongoing "processes". In this way, once-singular crafts can come together and act with each other, expanding their capacities and extending their abilities. As they do so, we are reminded of crafts' affordances as performative "doing things"; we can sense their capacities as verbs – of their ability to generate, to have effects, and prompt affects. The manner through which assembled crafts are performed generates different styles of endeavour. It manifests different idea(l)s and visions and can results in different "things" being produced through its practice. Craft is therefore a dynamic, responsive, mutually informing circuit of things and skills, of combined and constitutive products and processes, which come together to define our world.

In relation to surfing spaces, craft are at once the things that surfers assemble with to merge with the breaking wave, but it is also the ways in which they do so. Craft then refers to the glide, slide, or otherwise ride surfers choose to perform. This amalgam of noun and verb, of process and performance, in other words this assemblage of craft, produces the cultural geographies of surfing spaces.

With the diversity of surf-riding craft at their disposal (see Chapter Five), some switch between technologies dependent upon conditions and mood. As surf-rider Mike Lay states with regard to his own position:

> I hav[e] an open mind [about craft] and rid[e] whatever I want! ...I like getting into waves early, I like glide, not having to wrestle a board to make it do what you want. I surf a thruster every now and again and always enjoy it... I seriously cannot remember the last truly bad surf I had. I think that is down to riding a log when it's small, riding a shorter board when it's bigger or hollower... I used to ride high performance and loved it, but now I ride logs and love it even more. The bottom line is happiness, if people are happy doing what they're doing then rad
>
> (in Pumphrey, 2014: 43).

As Lay's words suggest, different craft work in different conditions, and their names indicate a flavour of their crafting style; yet whilst some are happy to retain versatility and conviviality with respect to craft choice, others are loyal to and defensive of one particular technology. As we will see, since the late 1960s, one type of surf-riding craft has been scripted to be the most appropriate version of the geo-cyborg assemblage, and this code of craft has been underpinned by a politics of verticality.

Dominant crafting: A politics of verticality

> As the lull drags on at Second Peak and the sun slowly burns off the June haze, the silence of the line-up is broken when a pair of dolphins burst into the air a few yards away, inside the kelp beds strung across Pleasure Point. Heads swivel again as a sea otter pops up in the dolphins' wake and backstrokes its way past groms, long-in-the-tooth long-boarders and twenty-something rippers. At last a set rolls through and I catch a glassy right toward the cliff-side house that Jack O'Neill built. There's a timelessness captured in these Santa Cruz moments… the surfer's up-close-and-personal encounter with marine mammals that have cruised these waters for millennia, the sandstone bluffs and the redwood studded hills… these waves too, first ridden 129 years ago when three visiting Hawai'ian princes fashioned boards from redwood planks and hit the surf near Steamer Lane.
>
> There's something else that has remained the same over the past half century, the image of the surfer. Whether appearing in surf magazines and videos or occasionally dropping into the mainstream media, the surfer invariably is presented as young, male, white, and obsessed with speed
>
> (Woody, 2015: 70).

This quotation from Woody highlights a number of key facets of the surfing spaces script and its constituent codes. It places us in a littoral location, specifically California, a picturesque, cinematic, American "frontier". It reminds us of the role of media architecture, of commodification, of surf-entrepreneurs in imagineering the script (Jack O'Neill's company, for example, self-celebrates 60 (+) years of innovation, see http://www.spatialmanifesto.com/research-projects/surfing-places/on-trend-and-on-the-wave/cultural-authenticity). It pays passing reference to the diversity of surf-riding (acknowledging groms and long-boarders), but ultimately focuses attention on one particular geo-cyborg (the young, male, white, board-rider). A surf-rider who performs a particular form of "craft": a short-board, at speed. In Woody's words, we are reminded that when the word "surfing" is considered, particularly for the uninitiated, it is this combination of craft that is called to mind. This is the craft that sold the magazines and built the mansion, this is the craft that fuelled the success of the surfing spaces script. It is this craft that has come to colonise surfing spaces.

> It seemed to me the lamest aspect of this [surfing spaces script] was that it focused on such tiny differences. Line-ups hen-peck those who don't conform
>
> (Taylor, 2019: 106).

> I noticed graffiti on a handrail overlooking the break that read, "No kooks or long-boarders, please". I've seen graffiti like this all over the world, but it took a New Englander to add the proper "please"
>
> (Taylor, 2005: 40/41).

Different geo-cyborg assemblages and the codes of craft that constitute them become one of the "tiny differences" (after Taylor above) that (b)order line-ups all over the world. Those that are perceived to be non-conforming to the dominant codes are marginalised and considered out of place (see Chapter Thirteen). This process of categorisation and division between surfing ontologies imposes a modern atomisation and hierarchy onto surfing spaces which runs counter to many of the relational sensibilities which mobilise most surf-riders' involvement in the first place (see Chapter Six). These codes are enacted to (b) order surfing spaces and gain, then retain, advantage for particular surf-riding geo-cyborgs; and this chapter argues that this code of craft can be understood as implementing a "politics of verticality".

The politics of verticality (to apply Weizman's phrase, see Elden, 2013: 37, in a new context) imposes two key dimensions to the code of craft. Firstly, it prioritises those who stand-up (i.e. those who choose to be vertical) to ride, and secondly, it valorises craft which can (vertically) ride up and down waves (and not "simply" along). We will critically engage with each of these dimensions of the politics of verticality in turn.

Stand-up to be counted

The first dimension of this politics of verticality is the valorisation of those who remain upright to surf-ride. From the perspective of this code, surf-craft involves the use of a surf-board, and specifically, the act of standing on this board. Those who fail to do so, in other words those surf-riders who may prefer to lie, kneel, or sit in order to craft their way through a wave, are rendered marginal by this code, as are the material crafts which are used in this process (including the human body itself, body-boards, kayaks, and surf-skis).

Concomitantly, the relational sensibilities that are produced through these geo-cyborg assemblages are also ignored and silenced. These may relate to the craft of the kayak[1], or the body-board, as Taylor sums up in relation to the latter:

> Bodyboarders like to say that they perform their craft "in" the water, and not "on top of it". This gives them a more connected, nuanced understanding of the ocean/land interface. [As a result] the experience of riding a bodyboard [is so different] that many stand-up surfers may never fully understand [it]. [As Mike Stewart, 9 times bodyboard world champion and 14 times bodysurfing world champion puts it:] "it is such an intimate experience, a different view, you're very connected, you see what the lip is doing at all times… you're so far back there"
>
> (2019: 105).

The lack of awareness about the relational sensibilities of alternative craft can lead to a concomitant ignorance of how these experiences inform the language of surfing spaces more broadly. As Taylor identifies, the terms, "pits, slabs, trenches, riding the foam ball, shocky jockey. All of these were coined by bodyboarders" (ibid.). Cumulatively, such marginalisation can lead to assumptions about the relative importance of these alternative crafting skills in the evolution of surfing itself; for example, bodyboarding is often labelled as "an adaption of stand-up surfing" (Warshaw, cited in Waitt and Clifton, 2013: 488), rather than co-existing in many endogenous surfing spaces (and it being most likely that stand-up surfing is an extension of bodyboarding, rather than the other way around). This marginalisation is confirmed when such pursuits are dismissed as pastimes for children or novices, rather than serious craft for authentic surf riders (see Waitt and Clifton, 2013, for example, and Chapter Five[2]).

> 'You can't talk about bodyboarding without talking about the struggle, especially in Southern California' said San Diego bodyboarder Casey Allred. 'For an average sport, there is big-time social struggle to it'
>
> (in Taylor, 2019: 94).

As important as it is to acknowledge the language of surfing spaces that has originated from the performance of alternative craft, it is also vital to call out the language articulated by some surf-riders to diminish these alternative craft and their users. In a similar way that geographical outsiders are caricatured and marginalised by some local surf-riders, those using alternative craft are also labelled as "strange(rs)" (see Chapter Thirteen) and face the prejudice of the dominant. As Taylor puts it:

> Dick-draggers, boogers, half-men, lid-riders, spongers… these terms used to be everyday epithets stand-up practitioners applied to surfers who paddled out on flexible hunks of packing foam. Traditionalists verbally harassed, dropped in on, and did their best to intimidate bodyboarders. Through the 1980s and 1990s, this bias was taken for granted, like something handed to you with the purchase of a leash or a bar of wax. South African bodyboarder Boots Crossley summed it up as, "you don't stand, so you are less"
>
> (2019: 94).

Based on an ignorance of the craft and a wish to maintain their own advantage (and perhaps the perceived threat posed by the practice of these technologies as we will discuss further below), the "sophomore" racisms of the surfing spaces script (see Chapter Ten) here become expanded to include the dismissals of different craft through other phobia. In this first aspect of the politics of verticality we can identify a toxic mix of patriarchy, phallocentrism, and homophobia; as Evers confides:

> I got to hang out with crew who were dickheads, who set strict rules that narrowly defined manhood and privileged one way of being a man at the expense of all others. It was a model of man hood that excluded… blokes from backgrounds different to ours
>
> (Evers, 2010: 3).

And as Ormrod points out, these "strict rules" meant that,

> ... kneeboarders were often dubbed "cripples", and audiences at pure surf movies often yelled 'Stand-up you pussy!' when a kneeboarder was shown, merely reinforc[ing] antagonism on the waves between board riders and [others]
>
> (2007: 100).

In the media architecture that supports the dominant surfing spaces script such prejudice was rarely condemned. Articles and features did little to challenge the notion that not-standing up to surf meant you suffered from the "pussies' disease" (short-boarder Joel Tudor, in Mondy, 2014: 67). The injection of these toxic and prejudicial idea(l)s of gender and sexual orientation became so normalised that, as lisahunter puts it, it became difficult for "other sexual identities to explicitly exist" in many surfing spaces (2015: 179)[3]; as lisahunter goes on:

> many seem unaware of their cishet privilege. [...] The doxa of compulsory heterosexuality, heteronormativity, and cisgender privilege have made it difficult to recognise sex, gender and sexual[...] diversity and fluidity [in surfing spaces]. This privilege and doxa is often hidden in an unawareness of privilege that influences who is recognisable, valued, and positioned more or less powerfully in influencing the social field and how they participate in the field of surfing
>
> (2015: 173)[4]

Move along old timer, we're moving up

As we have seen, the politics of verticality includes prejudice against those who do not stand to surf, but it also includes similar antipathy against those who do not stand in the "right" way. This right way, in short, is riding short-boards.

> The opening volley of the "Shortboard Revolution" [...] was served on October 2, 1966, in Ocean Beach, San Diego. It was there, under overcast skies in small, soft surf, that Nat Young won the World Surfing Championships with a radical, hard-turning approach that stunned the ruling elite. When David Nuuhiwa, the lithe Hawai'ian favorite famed for long, elegant noserides, insisted, "You must try to blend into the wave," the cocky Young retorted, "I don't want to blend with anything!". Young's rhetoric was as inflammatory as his equipment
>
> (Barilotti, 2013: 17).

As we have seen in Chapter Five, in the late 60s technological developments enabled surf-boards to made more quickly, cheaply, and of lighter materials. This new materiality was mirrored in the new styles of riding that were introduced; lighter and shorter boards transferred greater speed and agility to surf-riders meaning you no longer had to go along the wave, but could go down, and up it too. Short-boards encouraged fast moves, cutbacks, and even aerial manoeuvres (with many moves similar to those undertaken on skateboard or snowboard pipes), in sum, short-boarders went vertical.

> We're not hunting perfect barrels. As long as we get a little beachbreak with a little ramp, we're gonna have a good time
> (Kai Neville, surf-rider and film maker, in Maxam, 2011: 76).

These new craft could achieve stoke for surf-riders in different types of waves. Short-boarders became interested in waves that broke in shapes similar to the quarter- or half-pipe, rather than becoming solely motivated by the (full) tube (as Neville notes above). Short-board craft excelled in their capacity to ride verticals, going up and aerial, often moving against the energy of the wave rather than with it, and transforming the wave into a ramp. Verticals became the gold standard; a master at (and on) this craft could pay service to the lip; they could: "turn straight up the wave… the idea [being] to go up as vertical as possible and turn as quickly as possible" (Cornwall Guide, 2021: no page).

> The modern short-board… is like a Formula One racing car. It's no wonder that the level of high performance surfing is what it is today – look what they're driving
> (Lopez, 2001: 25).

The short-board craft had the consequence of introducing a new kind of "high performance" to surfing. The most lucrative currency became the ability to perform strong, aggressive, and controlled moves, all at speed. As outlined in the incident with Young and Nuuhiwa above, to some extent, this favoured younger, fitter athletes, and marginalised any craft or rider who favoured "displays [of] slower movements, grace, and elegance" (Gibson and Warren, 2016: 23). As Gibson and Warren sum up, "When the short-board revolution… began, a dominant masculine style of riding became established… To surf [outside] this style was to shamefully 'surf like a chick'" (ibid.: 22).

This double edged "performance capital" (Ford and Brown, 2006), was (and is) celebrated both in media architecture and broader surf merchandising. A recent article in a UK surf magazine, for example, cheerleads the butchery of, "Ryan Callinan [as he] splices the guts out of an emerald Indo runner with searing precision" (Gartside, 2019: 18), whilst entrepreneurs continue to market their products using military, aggressive, even sexualised metaphors, as the following examples illustrate:

> The new fusion 3q zipper with over-locking teeth completely blocks out water. The reduced zipper length allows for more freedom of movement. Easy in easy out. Formula for aggression
> (O'Neill, wetsuit, advertisement placed in *The Surfer's Path*, 30, April-May, 2002).

> Fighting the cold war. Razor. Zone. Bullet. 5x3 wetsuit technology from C-skins.
> Put the boot in! C-skin boots and reef slippers... designed for aggressive surfing. [...] Push harder!
> (C-Skins, wetsuits, advertisement placed in *The Surfer's Path*, 28, Dec-Jan, 2002).

> Designed to fit. Made to grip. So you rip
> (C-Skins, boots, advertisement placed in *The Surfer's Path*, 64, Jan-Feb, 2008).

Echoing through the military metaphors and the tendency of surf companies to "weaponise" their merchandise, these adverts seek to service the domination of aggressive short-board riding over other craft. These companies' campaigns can be seen to, if not wholly create, then consolidate and perpetuate a "norm" that this form of riding is the orthodox and "proper" way to perform surfing spaces (Bourdieu, 1977, 1991).

Indeed, the politics of verticality which underpins the code of craft can also be seen to be active in other codes which (b)order surfing spaces. For example, it is possible to suggest that the code of "positioning" discussed in Chapter Twelve came into being as a direct consequence of the short-board. As short-boards have a small planning surface, they need to extract the maximum speed from the wave in order to enable their capacities. As Kotler suggests, this means that to catch a wave on a short-board "requires taking off later" (Kotler, 2006: 61). Premium is thus put on being positioned at the lip of the breaking wave (where speed and height are optimal). Similarly, due to the potential of the short-board to be manoeuvred in fast, flighty ways, space on the wave is now all-important. The safe existence of other craft on a wave nullify these capacities and thus it is possible to suggest that the code of "one surfer one wave" was scripted to service the short-board and its users[5]. Relatedly, it is also possible to suggest that the code of surf-rider provenance is a further consequence of this craft. The popularity of the short-board and its associated style raised the numbers of young men in surfing space hotspots, but as it did so, it also minimised the carrying capacity of each wave. As Waitt and Frazer comment, "historians and sociologists have framed localism around the practices, motives and experiences of surfers who ride *short-boards*" (2012: 327), suggesting that other craft (in both its senses) may realise in practice different attitudes towards the importance of provenance.

The "malice" of "the dominant culture"

It is possible to identify, therefore, that the politics of verticality enabled the short-board craft to achieve dominance in surfing spaces. Once gained, however, this dominance was actively retained with respect to other craft. The rise of stand-up paddle boards in the twenty-first century, for example, posed a threat to the hard-won dominance of the short-board. Stand-up paddle boards enable their riders to be vertical whilst they paddle for optimal location at the leading edge of a wave, and as a result, to identify this ideal spot earlier, and due to their use of a paddle, to get to this spot more quickly. As a consequence, SUPs present their riders with competitive advantage over other craft, including short-boarders, and resultantly pose a threat to the (b)orders established by codes of positioning and provenance. In a similar way, long-boarders also enjoy competitive advantage in some waves. As Waitt and Frazer state, "long-boarders can… ride a much smaller, less powerful swell" than short-boarders (2012: 330) whilst they also have more time to take off for waves due to the larger planning surface of their craft; as a consequence, both paddle- and long-boarders are able to "'control' a surf-break by being positioned further out in the ocean" that other craft (ibid.). If these craft were numerous at any surfing space, stand-up short-boarders would be relegated from "top dogs" by these "top logs". In the face of such potential threat, many short-boarders continue to assert their assumed privilege in surfing spaces.

> … the malice that the dominant culture had once held for bodyboarders migrated on to stand-up paddle-boarders (terms emerged for them: sea sweepers, custodians).
>
> (Taylor, 2019: 94).

As Australian surf-rider Harrison Roach puts it, there continues to be:

> a backlash against loggers and plenty of aggressive short-boarders willing to vocalize it. "I'd be surfing Tea Tree [near Noosa, Australia], like two foot, offshore, absolutely perfect for logging and the… guys would say, "What the fuck you riding that thing for Harrison? I thought you'd quit that crap." Then they would bounce down the line like fools while I was having the time of my life. It was ridiculous
>
> (in Mondy, 2014: 68),

And as surf-journalist Jarvis states,

> There are people out there – good, solid, intelligent people – who refuse to ever ride a long board. These people believe that long-boarding is a sign of weakness, of age. That you've lost the war against vigour and youth. … Each to their own, but that's just one approach, one mind-set. There is another approach that dictates that a surf is a surf, it's fun and healthy, and it doesn't matter what equipment your riding own, from a short- to a long-board to a Stand-up Paddleboard. After many years of being a full member of the elitist short board group, I jumped ship, crossed the floor and embraced the second mindset – the "ride whatever" group, the swingers of surfing fraternity
>
> (Jarvis, 2014: 66).

Conclusion

As Ford and Brown (2006) confirm, in recent years – and perhaps since the "short-board revolution" of 1968–1970 (see Kampion and Brown, 2003) – short-board riders have accorded themselves elevated status and respect in the surfing field. In turn condoned and celebrated by the dominant surf media, this code of craft has brought with it certain power dynamics which favour young, athletic, male surfers within surf culture, and serves to (further) marginalise everyone else. Augmenting the protectionist tactics used by locals (as we have seen in Chapter Thirteen), this group adopts a range of derogatory labels which "insist" that those who do not craft the waves in their way are somehow "deficient" in comparison (see Booth, in Waitt and Clifton, 2013), and subsequently beneath them in their own imagined/imagineered surfing space hierarchy. The success of these tactics has led to the imposition and perpetuation of this code, to the point at which it has attained hegemonic status, as Mihi states, it is now the "commonly accepted norm in and outside of the surf" (2015: 91).

It is in this dominant reality that the experience of Kimball Taylor at the fabled and often extremely challenging Riley's surf break in littoral western Ireland stands out (or is rather a salutory out-lier). As this chapter concludes, it serves to raise questions that critique the dominant ways in which surfing spaces have been scripted and coded, influence by the Californian imagineering process. It at once demonstrates the common acceptance of the short-board code of craft (Taylor acknowledges how rare it was for him to witness a gathering of bodyboarders), whilst also offering a more convivial and inclusive example of how it is possible for a range of craft to co-exist in surfing spaces more generally:

> the days I visited Riley's [...] of the nine or so surfers out on any given day, only two or three would have been stand-up surfers. The rest rode bodyboards. The fact that the former group didn't stand didn't mean it was easier, only that they were able to surf places on the wave that stand-up surfers couldn't.
>
> Fergal Smith, a professional stand-up surfer who had his pick of any wave on any day at Riley's... did something that no self-respecting California professional surfer would do. He picked up a bodyboard, and began to catch waves prone. It was swapping a fork for a spoon when the stew was that good. The scene sparked a number of insights for me. One, that I hadn't witnessed such a vibrant gathering of boogers, well, ever, and that actually, I hadn't set eyes on more than one lid-rider at a time in years. Two, that this was the kind of place they assembled... at a wave that so treacherous, few stand-up surfers cared to mess with it. And three... that malice between stand-up surfers and bodyboarders was not universal, or even rational. Everyone present felt so lucky to even exist then and there the tiny differences that defined others simply fell away.
>
> That altered landscape begged questions I should have asked long ago: What, exactly was it that fed the malice stand-up surfers held for

bodyboarders? How was it that the simple choice of an implement used in a shared hobby had become definitive and conflict laden?

(Taylor, 2019: 96–98).

Despite the hegemony of the code of craft, this example suggests that in various surfing spaces around the globe, surf-riders choose to perform their own cultural geographies before and beyond the dominant script. As we will see in Chapter Fifteen, such transgressions can also be seen in relation to the hierarchy of surf-riders based on gender. Before we explore these other practices, we first engage with the ways in which the dominant script has created codes which serve the interests of (all those at Taylor's experience of Riley's:) male surf-riders.

Notes

1 Surf kayakers, like bodyboarders, sit low in the water, "and are always partially immersed" (Sanford, 2007: 882). Surf-skis are thus the closest technology to surf-boards as they are flat(-ter) boats on which you sit, strapped in, often around your waist or over your thighs. When on a surf-ski it is as if you are sitting on a narrow, precarious tray. As a consequence, a surf-ski is liable to tip when stationary, but will glide effortlessly when propelled by the sea's momentum (or your paddling) (see centralcoasttoday, 2008) http://www.youtube.com/watch?v=VClP4BclJic for an example). Nevertheless, the few inches of submergence created by kayakers, body-boarders and to some extent surf-skiers, means that they cross the boundary between air and depth, and transgress this surface with their craft. As surfboards have no depth, no edges or rails (that is, the side sections of the hull of the kayak) and as a consequence surfers experience a *flatter* sense of buoyancy.
2 Waitt and Clifton summarise the dominant view as follows, "Requiring little bodily strength, skills or expense, the Morey Boogie board was immediately embraced by thousands of children as play. Boogieboarding became an integral part of family beach cultures and continues to enthral thousands of young beachgoers globally" (2013: 488).
3 Indeed, as Taylor (2019) argues, bodyboarding in California effectively ceased to exist due to this prejudicial coding.
4 Or as Evers puts it more succinctly, "Homophobia is rife in the surfing culture" (2006: 4).
5 As Finnegan explains, "With long-boards, it had been possible, at least in theory, for a small number of surfers to share a wave. The frantic, quick-turning style required by shortboards, their need to he always in or very near the breaking part of the wave, meant that there was really room for only one guy on a wave now" (2015: 88).

15 Codes of Gender

Introduction

As we have seen, although the surfing spaces script suggests that the breaking wave is a place akin to an open frontier, in practice littoral locations are (b) ordered by a range of codes. In this chapter we explore the third in the set of codes which orient around the positionality of the surf-rider, shaping the spaces of surfing in terms of gender.

As we have seen in Chapter Ten, the imagineers of the 1950s and 60s framed the surf-rider as a heroic protagonist who, through his brave and dramatic encounters with surfing space, was able to escape the confines of mainstream culture. The cultures and geographies of surf-riding were all about, "the man on the wave, and …the rest is bullshit" (Jeff Divine, Research Interview). The use of the male pronoun here is no accident, it directly reflects the masculine bias of the surfing spaces script as it was initially imagineered and subsequently perpetuated into the twenty-first century. This surfing space, according to its script, was a culture and geography in which privilege and entitlement was claimed and enjoyed only by men. In this chapter we critically explore how this patriarchal vision of surf-riding was (and still is) "institutionalized" (to use the words of McDowell with reference to broader society (1999)) through its script, codes, and architecture to enhance the idea of the superiority of men, and in particular a specific type of masculinity. In this chapter we will identify how all bodies that do not conform with a dominant idea(l) of masculinity have been "abstracted" from the waves (see Chapter Five), and when women do attempt to re-claim a space within the littoral, the chapter examines the ways in which existing codes render them invisible, inferior, or only granted status by men, on masculine terms. In these ways, the role of woman-in-the-waves is coded in line with the broader "archetype" of the western surfer, as an object "out of place" on the waves, and subservient to the male surf-rider on the shore. Yet, as we will also see in this chapter, this diminishing and reductive role of woman-as-accessory is being challenged and successfully transgressed by a range of practices, including the establishment of media architecture that caters specifically for women surf-riders, and which introduces its own version of the surf-riding archetype. These moves, combined with the others which challenge the orthodox framing of surfing (and which are noted in

Chapters Twelve to Sixteen), call into question the viability of the dominant surfing spaces script in the twenty-first century (see Chapter Seventeen).

To (re)state an important position, just as all surfing spaces are relational and ongoing accomplishments, as are all humans. Not all men are the same, nor all women, and some human's gender compositions are not even encompassed by these two (no longer recognised to be) exclusive categories. When considering the (in)justices in relation to the (b)ordering of surfing spaces, it is important to retain nuance in order to respect the diversity of surf-riders included in these cultural geographies. Due to the diversity involved in these assemblages, it is possible for a range of apparently contradictory experiences to exist alongside one another in contemporary surfing spaces, with both corresponding to and calling back a "feel of truth"; thus when Australian surf-scholar Rebecca Olive points out that,

> Surfing in Australia is described as sexist and exclusive (Booth, 2001; Comer, 2010; Evers, 2004, 2010; Henderson, 2001.) … But my recreational surfing experiences are fun, pleasurable and inclusive
>
> (2013: no page).

it does not necessarily mean that either of these statements are incorrect, they both can co-exist as accurate reflections on a general state (the former) and specific assemblages (the latter) that have been imagineered or crafted to suitably reflect the idea(l)s of their respective participants. However, it is also clear that simply because these statements co-exist in the same sentence may not mean that their respective participants and idea(l)s are not in tension on some waves, and it is to these tensions that the chapter now turns.

"We asked the guys if we could surf their wave" (Taylor, 2005: 51)

> Women riding waves is nothing new. We've been right beside our surfing brothers since the first kids dragged fragments of coconut palm into the ocean and realised it was absurdly fun
>
> (Hill, 2020, 2021: 32).

As far as it is possible to determine, across the various endogenous histories of surfing spaces (see Chapter Nine), a range of genders have enjoyed the privilege of surf-riding. However, the imagineering of the surfing spaces script in the 1950s extended the patriarchal relations that (b)ordered terrestrial life in contemporary California (and elsewhere in the western world) into their idealisation of littoral locations. In line with the dominant culture of the times, surfing spaces came to celebrate "youthful white masculinity" (Dominguez Andersen, 2015: 511), with the idealised gender role typified by the "natural, rugged, and rebelliously masculine [arche]type" (ibid., see also Chapter Ten). The main media architecture (such as *Surfer* magazine, see Chapters Ten and Eleven) implicitly claimed to represent the whole of surfing culture (*Surfer*, for example, was not titled *Male Surfer*), yet focused almost exclusively on men. This assertion of control through omission reduced the capacity of women to be validated or recognised as appropriate surf-riders, and legitimated their further marginalisation by those emboldened by the script of surf-riding. As Stedman states, the media's "...silence [was] a form of discouragement, and the limited coverage of women in traditional male-dominated surfing magazines effectively denied them access to 'the symbolic resources needed to identify as surfers'" (1997: 86, cited in Booth, 2011: 13). As we have seen in the images presented in Chapter Ten, when women were included in surfing media, they were not presented as key protagonists in surfing space, their role was marginalised and diminished, becoming reduced to subservient onlookers and supporters of the male's role (see examples presented here http://www.spatialmanifesto.com/research-projects/surfing-spaces-gender-codes). This positioning not only implied specifically heterosexual relations between genders, but also more wide-reaching power relations, (b)ordering the waves as a space that belonged to men, and relegating women to the shore. Through reduction, objectification, and diminishment in words and images, the surfing spaces script compounded a broader system of disempowerment which gave women few real choices but to acquiesce to the roles assigned to them; as Rinehart puts it (when reflecting on the coding of women in surf movies);

> Passive females were presented as helpers to the men, as admirers of the men, as diminished beings whose reduced agency never came into question
>
> (2015: 558).

The sexism and misogyny inherent within the script was identifiable in those centres that imagineered the script. As seminal surf photographer Leroy Grannis put it in relation to California, the "perfect 'surfer girl' was expected to wait hours for her surf hero to come in" (2013: 117), whilst Pearson, in his work on surf media in Australia, notes how women were reduced to the role of "accessory" to the (male) surf-rider[1], as Tracks (an Australian surf magazine) prescribes:

> ...The surfing woman is meant to be a handbag, a decorative accessory for the shaggin' wagon like an expensive cassette recorder. For comfort stops who can cook, fuck and make your mates jealous by being as desirable as possible while remaining loyal and clean; a faithful dog who can do the shopping to keep the surfing body beautiful functioning in its holy watery glory
>
> (Tracks, 2: 51, cited in Pearson, 1982: 129).

By reducing women to "chicks", "dogs", and "vessels", one can see that this surf architecture exemplifies a perspective which scripts femininity and women as less than human, and (b)orders them to the margins (in this case, the place of the camper van or the beach). In this way the cultural preferences of bodily order are translated into material geographies: places are taken and made in line with prescribed yet prejudicial gender roles and relations.

In more contemporary surf media, these gender roles are just as identifiable (and a range of examples are available here http://www.spatialmanifesto.com/research-projects/surfing-spaces-gender-codes). Cohering with the subservient position of the western archetype which frames the overall script, these images serve to illustrate that the dominant media architecture still frame the beach to be the most appropriate location for women, marginal to where the surfing action is. Those women depicted waiting on the shore are young, slim, and often clothed in revealing swimwear (or none at all), their composition and framing selected by (male) editors to legitimate a gaze which objectifies their bodies and sexualises their form, often in ways that enhance their submissiveness to dominant hetero-fantasies; as Pepin Silva (a female surf photographer) states,

> I think a lot of men's photographs of women surfers are somewhat sexualised... it's the angles and the way that they are shot...
> (Research Interview).

so much so, that as Pepin Silva concludes, "over the years [women in surfing have] had their imagery sexualised to the point of being, in many women's eyes, soft core porn..." (Research Interview). Rinehart makes a similar point, confirming that in his view "the underlying sexism as the basis for male-driven, male-centered, male-viewed surf porn has only intensified" (2015: 559).

Established on land, replicated in the early imagineering of the surfing spaces script, and perpetuated by ongoing media architecture, patriarchal relations became dominant within the majority of surfing spaces. These prejudicial idea(l)s consolidate the homosocial "fratriarchies" (after Booth, 2004: 99) which we will see influence the development of the surfing space script in relation to travel (Chapter Sixteen), and exploit them to further extend their privileges in littoral locations. Indeed, for those socio-demographics which by lottery of birth meant they could be considered legitimate in this version of surfing spaces (i.e. male, young, able-bodied, and athletic), these fratriarchies actively police their (b)orders by not only, "jealously protect[ing] their territory against outsiders" (ibid, cited in Beaumont and Brown, 2016: 280, see also Chapter Thirteen) but also ensuring that no progressive "trojan horses" could destabilise their "rights" from the inside; as Clifton Evers recalls from his own experience within one such group of locals:

> Nobody ever told me it was actually OK to question the model of manhood I was being shown, even when it didn't seem right. If I did ask questions, I ran the risk of being beaten up, paid out on, bullied, yelled at to "toe the line", or excluded from the crew
> (2010: 3).

Through such external and internal (b)ordering tactics, codes of gender become,

> so entrenched and implicit that those who fit them best are able to manipulate people outside of the norm to become active and complicit in

regulating their own behaviour. This in turn validates the dominance of the most powerful group (in this case the male majority), continuing to locate existing norms as "authentic"

(Olive et al, 2015: 270).

In this prescription and subsequent normalisation of prejudice, the cultural geographies of surfing spaces became coded with patriarchy, in Rinehart's words, "the lesson is that surfing has been fundamentally linked to and for males, and many reinforcers of this ideology reproduce the inequities" (2015: 559). Far from being a space of freedom, the codes of gender mean that if you want to go surf-riding, you have to ask the "guys" if you can surf "their" waves (after Taylor above).

Finding space to surf within dominant (b)orders

Due to the hegemony of this surfing space gender code, any feminine or female bodies who wish to surf have to do so in the face of prejudice. This could be experienced in a range of ways; from having to engage with presumptions about insight and expertise (as males may assume superiority and feel the need to patronisingly "man" plain craft-use to females (see Bush, 2016: 300), or respond to assertions that women's mental and physical capacities mean they simply could never surf in the same way (i.e. as well) as men. This point is made clear by big-wave surf-rider Buzzy Trent who suggests that,

> ...girls are much more emotional than men and therefore have a greater tendency to panic. And panic can be extremely dangerous in... surf. Girls are weaker than men and have a lesser chance for survival in giant wipeouts
>
> (in Booth, 2001: 3).

Due to these essentialist expectations and assumptions, it is common for female surf-riders to face (more) complex cultural obstacles (than men) when entering any surfing space. Just as (male) novices will need to execute their skills without error on their first wave in a littoral location (see Chapter Thirteen), any female surf-rider may feel this expectation keenly, as a respondent in Comley's research exemplifies:

> You have to get that first wave because everyone is watching you and waiting to see if you can perform. You have to represent, especially as a woman. You have to show them (men) that you deserve to be out there. The guys don't have to do that and we (women) know that, but we keep doing it anyway
>
> (Katie, 44, intermediate, in Comley, 2016: 1294).

It is also possible for female surf-riders to be simply ignored or openly dis-"respected" on the waves[2]. UK surf scholar Stephanie Merchant identifies, for example, that:

> intermediate and advanced women are dropped in on considerably (30%) more often than men of the same standard... as Chloe considers; *"they [guys] always snake me, and then I put pressure on myself, cos I feel like I can't fall off the wave even though it's mine to do with as I feel, but then the guys think I'm an idiot when I fall off, like I've wasted it* [the wave]"
>
> (no date: 43/44, emphasis in original).

Ironically perhaps, competency in wave-riding heightens the possibilities that female surf-riders will face intolerance and prejudice from men. In these circumstances, some male surf-riders feel entitled to flout the code of positioning and drop-in on female surf-riders, in effect treating their bodies and expertise as threatening to the dominant code. Thus when engaging in surfing spaces, female surf-riders have to face a range of, "cultural understandings and expectations of the ways that activities should be performed, or assumptions about male and female performances" (after Olive et al, 2015: 261). Such codes of gender positionality threaten to impose essentialising norms with respect to the culture of surfing craft and the geography of wave-riding.

"Girls are better off and look more feminine riding average size waves" (Buzzy Trent, cited in Booth, 2011: 3)

Codes of gender positionality translate into the construction of (b)orders of surf-riding based on femininity and masculinity. In line with Trent's pronouncements (above), (b)orders are constructed which prescribe that female surf-riders could not and should not surf big waves, and could and should look feminine when they surf. This construction suggests that the aesthetics of elegance and grace (which as Chapter Fourteen has already discussed means a rider "'shamefully' surfs like a chick") define what femininity means in practice: as a diverting, but less important craft when compared to men. Indeed, this reductive and relegating construction is the counterpoint to a particular version of the masculine ideal. Based on and extending the politics of verticality (see Chapter Fourteen) this construction suggests that men can and should go big (i.e. they are better surf-riders and better men if they ride "waves of consequence"), and on more regular waves, they should shred, cut, and gut at speed, in order to considered a good surf-rider[3]. Through these essentialising constructions, the presence of novice, intermediate, or expert women-on-the-waves can be made acceptable (to male-surfers), and the surf-entrepreneurs and media architecture that service them can give the appearance of inclusion without destabilising their dominant code and advantages[4].

De-coding gender

However, just as the surf-entrepreneurs of the 1950s and 60s identified, if and when different socio-demographics begin to participate in surfing spaces, new profits are possible. Indeed, the "'surf industry', 'surfing media', 'surfing industrial complex', and 'surfing culture industry'" (after Hough-Snee and Eastman, 2017: 85) which was promulgated due to the success of the initial surf-entrepreneurs, have identified the opportunity. As Bush states, in the 1990s "the surf industry finally recognised the lucrative new market at hand: the other half of the population" (Hill, 2020/21: 36).

In order to respond and capture the potential of a female surf-market, the media expanded its architecture to cater specifically for it. Rather than integrate women directly within existing titles,

> in the mid-1990s, the surf media began to produce new surfing magazines devoted entirely to women. In 1995, *Surfing* added *Surfing Girl* as an annual insert and in 2000 it produced six editions of the supplement
>
> (Booth, 2001:13).

Other titles such as *Wahine, Surf-life for Women, Surfer Girl, Jetty Girl, Curl,* and *Women's Surf Style Magazine* became available, and it is clear that these imagineered key points of congruence but also difference from existing (male) surfing space scripts. This media architecture did not opt to "feminise" or otherwise insert women into the male "western archetype" as employed in the dominant script. Rather, it can be argued that they turned back to endogenous surf-riding traditions, often preceding the initial colonisation and commodification of Hawai'ian surfing culture (as outlined in Chapter Nine), to impose their own framing of what surfing spaces should be. This is exemplified by the title of a key magazine *Wahine* (and this term remains in the surf-riding social movement although this title is no longer in production). Etymologically, *Wahine* referring to a Polynesian woman and originates from the Maori language (but outside New Zealand it is also traditionally used in Hawai'i and Tahiti). This symbolism of this title is clear; in its simplest and most common understanding, *wahine* refers to woman or women, as female and wife, and thus focuses directly on women who surf-ride. The term is commonly paired with the word *Mana'*, which together alludes to characteristics of authority, charisma, prestige, integrity, spiritual power, and strength – in other words, traits that are noticeably broader and more complex than the limited conceptualisation of "female" in the dominant script (i.e. by turns objectified, ignored, or associated purely with the aesthetics of elegance and grace). This broader, deeper, and more complicated notion of womanhood is extended through continued use of more endogenous Polynesian vocabulary in this new surf-riding media; as the publisher of the first *Wahine* magazine introduces:

Codes of Gender 189

With "aloha" I welcome you to *Wahine* magazine... a reflection of the *Wahine* spirit in us all

(Edwards, 1995: 2).

The use of "aloha" again re-locates the "spirit" of this script not in California or main-stream/-land America, but in Hawai'i (the official "Aloha" state). Beyond a simple greeting, aloha refers in this case not to the significance of co-defining histories which Ingersoll has told us are significant for many Polynesian surf-riding cultures (see Chapter Four[5]), but rather to a conceptualisation of the surf-riding cyborg who emphasises,

> the coordination of mind and heart within each person. It brings each person to the self. Each person must think and emote good feelings to others…

(State Symbols USA, no date).

The emphasis of this "aloha" state, positions humans not as self-determined, isolated individuals, but rather as co-constituted relational beings, defined directly through their co-composition in broader assemblages; as State Symbols USA explains,

> Aloha means mutual regard and affection and extends warmth in caring with no obligation in return. "Aloha" is the essence of relationships in which each person is important to every other person for collective existence

(ibid.).

Positioning the (female) surf-rider within this symbolic framework can be seen to legitimise and support a very different conceptualisation of the individual and their surfing spaces when compared to the dominant (male) script. Through emphasising the values of pride, strength, relationality, understanding, and collectivity, this media architecture is scripting a different world to that of the rugged individualist setting out to search for stoke in an "open frontier". For many, this alternative surfing spaces script was an important piece of imagineering, giving voice to the possibility that a different world to the dominant was possible; as surf-rider and -scholar Easkey Britton states,

> When I was a kid growing up on the remote northwest coast of Ireland there was little or no surf scene, and I relied on my annual birthday subscription to what was then the only women's surf publication, *Wahine* magazine. That was my way of connecting with what other women and girls were doing in surfing. One of the first issues I got had a beautiful image of a Hawaiian woman on the cover and inside was the story of a remarkable modern day surfing queen and water warrior battling breast cancer. Her name was Rell Sunn, called "Queen of Makaha" by her

friends. From that moment, she became my role model. It is no surprise that Rell's Hawaiian name, *Ka-polioka"ehukai,* means heart of the sea. She was a real embodiment of the aloha spirit of surfing and inspirational in terms of bringing kids into surfing and pushing the standards of women's surfing at a time when they were getting little or no support. Her philosophy of aloha and respect for the *mana* of the ocean has influenced my whole life

(2015: 119).

The alternative script emerging from the new media architecture also diverges from its dominant counterpart in terms of its inclusive approach to craft. In its framing, emphasis on the hierarchies between craft and riders are foregone and replaced with an approach which explicitly celebrates a mutual respect between technologies for oceanic engagement, stating for example that:

> Wahine is for all women who love the water, whether she sails, windsurfs, bodysurfs, kayaks, canoes, swims, bodyboards or just stands in appreciation at the water's edge, and particularly the surfer, who we honor in this premiere issue
>
> (Glazner, 1995: 3).

Although such a script could be argued to be a pragmatic attempt to maximise its circulation[6], this inclusivity nonetheless aligns constructively with its broader "aloha" philosophy and the relational "spirit" it imbues into its framing. Indeed, within the magazines themselves, a range of craft types enjoy equal billing (in the first issues of *Surf-life for Women,* for example, long-, short-, and body-surfing share space in same photo-spread (see https://issuu.com/historyofwomensurfing/docs/premierissue). These imagineers overtly call upon their readers to position themselves within broader networks of culture and environment (as Elizabeth Glazner, when Editor of *Wahine* states, she was "eager for a publication that is a tribute to women and water, in honor of the planet's life-giving elements and appreciation of the simple things that nurture us all" (1995: 3); and positing an attitude to travel that (in contrast to what we will explore in Chapter Sixteen), "isn't only about finding good waves; it's about meeting people, making friends, and experiencing the world beyond our backyard" (Makarow, 2002: 73)[7].

The script of surf-riding as imagineered through women's media architecture thus stands in clear contrast to that of the dominant male framing. In opposition to the western isolato-archetype (see Chapter Ten), this script seeks to generate a version of the surfing space assemblage which celebrates relational co-constitution, inclusivity, and a broader affective embodiment. One could argue, indeed, that this vision for surfing spaces is not in alignment with "western" cultural values at all, but has more in common with archetypes that are more often associated with "eastern" story-lines and frames; for as Storr suggests,

> whereas Westerners enjoy having accounts of individual struggle and victory beamed into their neural realms, Easterners take pleasure from the

narrative pursuit of harmony. ...In Eastern tales that [do] focus on an individual, the hero's status tend[s] to be earned in a suitably group-first way. [...] In Asia it's a person who sacrifices who becomes the hero, and takes care of the family and the community and the country. [...] For Westerners, reality is made up of individual pieces and parts... For Easterners, reality is a field of interconnected forces

(Storr, 2019: 83).

With this "eastern" frame of values in mind, when Glazner advises her readers to "seek balance" (Glazner, 1995: 12) this isn't simply an advocation of craft, but a call to temper obsessional aspects of stoke-searching in a broader awareness of other duties, commitments, and priorities. Indeed, when one engages directly with women's surf-riding, one can identify a cultural geography which reciprocally reflects and informs these alternative "eastern" framings. In female surfing spaces there appears to be less antipathy between craft (as Lisa Andersen (when World champion) states, "I think women should surf any way they want. It's a freedom thing in the water" (Glazner, 1995: 8)), a greater intrinsic respect for locals ("our community is one huge global family, connected through surfing" (Searle, 2015: 14), and a lack of selfishness and stoke-obsession in the water, as Bush suggests:

> ...women will let waves pass if they are deep in conversation, or will often give a wave to another woman. Women also state [...] that there is much less competitiveness between them. One respondent explained, "It's attitude. Men tend to surf defensively. Most of the women I surf with are laid back and comfortable with each other." Another stated, "Women are more helpful, less cocky and more sharing in the water than men"

(2016: 303).

Indeed, the preference for an alternative script to frame more feminine surfing spaces has led to the relational sensibility of "stoke" being differently defined within them. As Bush states, in the spirit of aloha, in these spaces stoke has "more to do with shared values than performance" (2016: 306).

Conclusion

The emergence of an alternative media oriented around and for women surf-riders has enabled the imagineering of a new surfing space script which both influences and reflects surfing spaces more generally. This alternative script is significant as prior to its emergence all surf-riders (both women and men) had to surf-ride within codes of gender imposed and perpetuated by the dominant western archetype. This code suggested that (amongst other values) selfishness, individualism, and aggression should be the natural hallmarks of these cultural geographies. Within this framing, any and all alternatives were marginalised and silenced, limiting the possibilities for "other" femininities (and non-dominant gender affiliations) in the water. As "Melissa" sums up from her experience, as a consequence of this code,

> There is no space for me in [that] line-up. I can fit by their rules, sure, but I can never say anything about them. ... This is what sexism in the water means...
>
> (in Robertson, 2019: no page).

Through this code of gender, "hegemonic masculinity [and] male surfers' way of being" (Franklin and Carpenter, 2019: 50) has been naturalised into the dominant version of the surfing space assemblage, and in turn this imagineering "acts to ensure male dominance and privilege" is maintained and consolidated (ibid.). Nevertheless, with the rise of women surf-riders, and the advent of a media architecture to serve them, a new set of values have come to script an alternative version of surfing spaces. The new framing can be argued to be thoroughly transgressive as it directly challenges the doxa of the dominant script. Firstly, it reveals this script to be political in nature, as just one way that surfing spaces could be, rather than the only way they must be. Secondly, it challenges the notion that only one type of "man" and one type of "masculinity" has the right to define what a good surfer looks like, what surfing culture might be, or what surfing spaces are; and thirdly, it offers a working alternative which creates the opportunity for all surf-riders to be different in the waves. Although this new script still imposes (b)orders on what surf-riding should be, its inclusiveness suggests there remains the possibility for codes of positioning, provenance, craft, and gender to be re-(b)ordered through practice, as Easkey Britton puts it:

> the willingness of the media to show that other possibilities exist is an important way to challenge and transform gender inequalities, creating space for new ways of doing [...] for men and women
>
> (2015: 125–6).

At this stage it remains unclear the degree to which this more "eastern" archetype can and will influence the dominant "western" script, and the nature of surfing spaces more broadly (but see Chapter Seventeen). It is possible to argue that in creating an alternative script through media architecture that runs parallel to the dominant, this script does little to directly challenge the values and frames of that media and its audience. Although this positioning may reflect the "aloha" spirit of its storyline, it nevertheless enables the "aggressive, powerful, and competitive" (Franklin and Carpenter, 2019: 50) codes of the western script to co-exist with it, rather than directly calling out its prejudices and contradictions[8]. Before we reflect on the potential influence of both these scripts for surfing spaces in the future (Chapter Seventeen) we turn to the final code in the set of surf-rider positioning which directly affects the nature and sustainability of surfing spaces across the world. In Chapter Sixteen we explore the code of the "trans-local surf-rider" to reveal further the prejudices and tensions at the heart of the dominant surfing spaces script.

Notes

1 As Pearson's research details, this editorial response from Tracks magazine was typical of the more tabloid surf magazines' unapologetically and systematically denigrating framing of women, "Some time ago in the letters column some brave soul pointed out that Tracks was profoundly sexist –like the surf scene it caters to.... What could you put instead of 'chicks'? Well, might one suggest 'lady' which is just a little better and of slightly less condescending nature? Point taken. All vessels shall hence forth be called ladies…" (Ed. Tracks, 53: 13, cited in Pearson, 1982: 129).

2 Stranger makes the persuasive case that women have been ignored by broader surf architecture by not making clothing that enabled them to surf; as he states:

"Cultural perceptions of women as ornaments who merely decorated the beach meant that manufacturers did not consider producing appropriate clothing for active female surfers. [...] Boardshorts specifically cut for the female figure solved the problem instantaneously. '"Now we can actually go surfing and not have to worry about fixing costumes', enthused Lisa Andersen, adding that old-style swimsuits had prevented her from performing 'radical manoeuvres'" (2011: 13).

3 As Stranger argues, this "essentialist approach of defining a risk oriented, powerful and aggressive style of surfing as masculine delegitimizes the experiences of the women who choose to find their joy in this approach, and equally, to define a graceful and rhythmic approach as feminine denies the male experiences (it also ignores the fact that this was a *dominant* style until the mid to late '60s)" (2011: 108).

4 An example of such essentialist and reductive inclusion is given by the article attributed to surf photographer Morgan Maasen, who wishes to celebrate women in the waves after he "found myself discovering the ethos of surfing …is… held in the hands of women surfing. The biggest air or longest barrel no longer captivate… my attention; it [i]s the dance of women on waves. I began to realize the pinnacle of surfing is expressed in style, and not maneuvers [sic]" (Maasen, 2019: 24).

5 And are notably side-lined by the dominant western script.

6 Indeed, with (male) magazines under pressure, many have turned to broaden the activities featured either on their pages or online, see for example theinertia.com

7 It is important to note that despite its progressive cultural and ecological leanings, this script remains tied into commoditisation which often plays in to notions of femininity which may be seen to be more stereotypical in nature; for example, *SurfGirl* offers "a kaleidoscope of inspirations, tips and ideas to make your life beautiful and shiny" (Searle, 2015: 14)).

8 It is important to note, however, that many within the surf movement do directly call out the patriarchy and prejudice existing in both the dominant surf media and in surfing spaces more broadly. In response to an article by surf-rider Emily Grimes, which suggested that "surf feminism" itself that was the problem in the water rather than the influence of the dominant surfing spaces script and its associated codes and architecture, "a collection of voices from the female surfing community" (including individuals associated with *The Institute for Women Surfers, Surf Senioritas,* and *Wave Wahines* (see http://www.instituteforwomensurfers.org/, https://www.facebook.com/surfsenioritas, and https://www.facebook.com/NDwavewahines/) offered their own views on what this chapter has called the codes of gender that (b)order surfing spaces. There follows excerpts from this dialogue:

"I am writing about what I see to be a misplaced, yet extremely prevalent phenomenon within our surfing culture, one that I worry might ironically discourage women from taking up surfing. Namely, the reams of articles written under the banner of Surfeminism, which claim surf media and culture are inherently hostile to women.

No one has ever told me to put my surfboard down. No one has ever told me to get out of the water. I have faced no barriers in learning to surf, other than those pertaining to the fact that surfing is quite hard, and also very dangerous, if you don't know what you're doing. No one has ever slapped my ass or made me feel uncomfortable in the water, solely based on my gender. Yes, I have been dropped in on. Yes, I have felt disrespected. Yes, I have been patronised, but for the most part I don't believe this is because I am a woman. I do not feel underrepresented by the media" (Grimes, 2019: 36).

"Women who ignore the experiences of other women do so because their privilege is working for them. Their privilege is keeping open their spot in the line up, sometimes at the expense of other women. If only the rest of us were so fortunate.

"Just because something doesn't or hasn't happened to you – doesn't mean it isn't happening to others right this moment on their local beach" – Yvette Curtis, Wave Wahines Founder.

Stories of women being targeted in the surf are frequent enough and geographically spread to prove that it's not a one off. The sea won't turn you away for being a woman, but the people in the line up sure will.

"As women, just pausing to think about how it felt when we first started to surf, when we first put on a wetsuit or our first solo surf at a new break can help us engage in feminism and to extend our privilege and experience to the women behind us, "Surfeminism" reminds me that I need to pay less attention to who claims rights to the inside and the set waves, and instead look back towards the shore and notice who has paddled out after me, and even who hasn't felt able to paddle out at all" – Dr. Rebecca Olive.

Our message should be loud and clear in and out of the water. The sea is here for everyone, regardless of sex, race, gender, sexuality, disability, class or beliefs. The ocean could be the great equaliser and a powerful tool for all those who need it" (Robertson, 2019: no page).

16 Codes of Travel

The "trans-local" surf-rider

Introduction

In this chapter we turn to the final code in the set of surf-rider positionality: the code of the "trans-local" surf-rider. As we have seen in Chapter Ten, the dominant script of surfing spaces extends the thrill and adventure of riding the breaking wave to include a range of associated activities, including surf-travel. Whether this wanderlust is simply from inland settlements to littoral locations, along the coastline to new breaks, or internationally to far-flung destinations, exploration has become intrinsic to the surfing spaces script. Yet as Anderson and Erskine argue, all mobility has consequences (2014). Travel has the effect of changing the individual undertaking it, whilst also bringing about a range of environmental, economic, and cultural impacts for the destinations visited. This chapter explores the consequences of the wanderlust within the surfing spaces script. Although we have seen in Chapter Fifteen how an alternative, "eastern", archetype of surf-riding may offer a more progressive approach to surf exploration, this chapter suggests that this is in stark contrast to the dominant approach and style of travel has been imagineered into surfing spaces, an approach which produces particular consequences for the individuals and destinations involved.

Firstly, the chapter will explore how the dominant script's framing of surf travel translates into a particular code of positionality. It argues that when an individual surf-rider goes mobile, the script encourages an identity change from a "geo-cyborg surf-rider" (see Chapter Five) to a "trans-local" surf-rider. Developing the example of colonisation (by state and capitalism) in Hawai'i (see Chapter Nine), the "exceedingly simplistic" approach to surf travel celebrated in *The Endless Summer* (after Laderman, 2014), and the implications of the western archetype that underpins their script, this version of the surf traveller no longer maintains respect for local provenance (as implicitly valorised in the code of surfer-positioning and explicitly so in its own code of -positionality), and wherever possible dismisses host populations as subservient to their stoke. Through celebrating and promoting a colonising approach to surf travel, the trans-local surf-rider assumes entitlement to (b)order destination littorals with the interests and values they bring with them. Such an approach either forces local surf-riders to adopt protectionist measures against them (as we have seen in Chapter Thirteen) or surrender their cultural

DOI: 10.4324/9781315725673-16

geographies to the extension of this script's colonising frontier. The chapter documents some of the cultural impacts that this "trans-local" surf-rider code has on destinations visited, before identifying some of ways in which a growing number of surf-riders challenge both the framing and consequences of this "trans-local" surf-rider code.

Scripting travel as exotic exploration

Travel has become central to the surfing spaces script. As few surf-riders are able to actually live in a littoral location (in other words, directly on a coast), especially in one that benefits from year-round, quality surf, all surf-riders at one time or another must travel to some degree to get their stoke. As we have seen, the surfing spaces script actively encourages travel, not simply by framing littoral locations as free and open utopia in which surf-riders can escape, but showcasing the opportunities to surf when they get there.

In every surfing magazine, adverts exist for regional, national, and international surf trips, whilst world tour competitions are championed, "new" surf spots introduced, and exotic locations celebrated. Surf films highlight the range of breaks available across the world (see Conroy 2009) and encourage local surfers to plan their own "odyssey" to find them (see Anderson, 2007). Along with advances in air travel, satellite communication, and surf forecasting, surf-riders are able to keep track of hurricane events and related swells around the world (for example online sites like "Magic Seaweed" offer comprehensive wave forecasts, allowing anyone to know where surf will happen regardless of local knowledge or expertise). Taken together, the surfing spaces script directly encourages the surf-rider to be a highly mobile modern nomad, living the dream of their own "endless summer", or at least endless waves (see, for example, Ouhilal 2011[1]).

Yet (surf) travel produces consequences, both for the locations visited and the individuals going mobile. When moving from one location to another, the person-place relations which come to co-constitute an individual (see Chapter Four) become "disoriented" (see Anderson, 2015), and opportunities arise to re-align identities and practices in new ways (see Anderson and Erskine, 2014). Individuals may travel for the opportunity to learn about other cultures, to broaden their horizons about different ways of living, or simply to liberate themselves from their own location for a short period.

In encouraging surf-riders to enjoy their own *Endless Summer*, the surfing spaces script not only suggests surfers should travel, but they should perform their travel in a particular way. From Hume's commodification of Hawai'i, to Brown's "surfari" films, surfing's media architecture has promoted an idealised form of travel, oriented directly to unspoilt surfing littorals; as Barilotti sums up, the dominant script suggests that,

> most surfers [should] travel not to experience another culture, but to find waves similar to their home breaks but without the crowds
>
> (2002: 93).

Imagineers have thus scripted surf travel to be all about the travails of finding and riding "new" waves. The ultimate "dream" of surf travel (following, for example, Preston-Whyte, 2001; Ford and Brown, 2006; Ponting, 2009; Laderman, 2014) is the discovery of a consistently breaking wave, and an absence of other surfers to spoil the experience. As Rae sums up (as he reminisces upon his travels in Bali in the early 1970s), this script has,

> always [provided] that vision: no traffic, no tourists, a raw coral track leading down to perfect waves… [this is] Surfer's Disneyland. [This script] had everything I treasured, in abundance. …Consistent swell, offshore winds every day, [and] no crowds
>
> (Jarratt, 2020: 126),

And as Ponting confirms, the

> imagery of perfect uncrowded surf in paradisaical tropical destinations has been the dominant theme in the surf media since its inception
>
> (2009: 175).

Such scripting of surf travel resonates with the colonial imaginary of the western archetype (see Chapter Ten), framing the world and its cultures as open spaces passively waiting for surf-riders' discovery. Despite the world becoming increasingly connected, this imaginary remains albeit reinvented in new ways, chartered boat-trips in South East Asia, for example, now enable modern surf-riders to become the "equivalent of 19th century "Gentlemen Adventurers"" (after Barilotti, 2002: 36, cited in Ponting, et al, 2005: 141). To this end, the dominant surf media continues to actively encourage surf-riders to, "go remote – as far off the beaten track as possible" (Matt, in Carve Magazine, 2014: no page) in their search for well-disposed surfing littorals without the endogenous surfers to inconvenience them. In this way, the media encourages surf-riders to nostalgically enjoy their own colonial adventure, gaining their sense of what it was like to be not simply like Rae (above), but also Captain James Cook, Jack London (see Part I), or Bruce Brown (see Chapter Ten):

> The further you head away from [development] the lower the crowd density becomes, and the closer you can get to the real [experience] of yesteryear. That deserted tropical island and those empty spitting waves are still there to be found; you just have to want them enough. Ever thought about chartering a seaplane?
>
> (ibid.)

Shedding skins: From geo-cyborg to "trans-local" surf-rider

Under the influence of this script of surf travel, the surfer who at home may be assembled closely with their littoral location (see Chapter Three and Four) is fundamentally disoriented. As they become mobile, the seduction of the script cleaves their ties to place and dissolves their geo-cyborg relations. As this occurs, the mobile surfers' field of vision narrows and a new identity is formed that is no longer oriented or "embedded" in around real places (after Ponting, 2009); the code of surf travel becomes all about living the dream(script).

The script of surf-travel thus transforms the surf-rider into a "trans-local" being. As Mandaville (1999) has noted, "trans-local" identities and practices are formed when people become mobile and take with them global, often rootless, cultural ideas and lifestyles. The script of surf-travel can be understood as one such "trans-local" vision. As we have seen in Chapter Eleven, this script actively avoided any "confrontation with the realities of political economy and the circumstances of global power" of which it was a part (after Routledge, 1997: 360) preferring to imagineer an illusion in which the surf-rider could intentionally extract themselves from their defining spatio-temporal relations and enjoy a surf-travel to fantasy frontiers. Through celebrating and promoting this (dream)-script, this dominant framing extends the "simplistic imagination" exemplified in *The Endless Summer* (see Chapter Ten) into a functioning "global" or "trans-local" code. This code validates an imaginary version of littoral destinations and silences inconvenient networks of responsibility that define their real-life counterparts. As West notes, this code discourages surf-riders from being concerned with these networks, so much so that they,

> literally do not see places, peoples, and events that sit outside of the fantasy. This is in part because of the ways that tourism operators structure experience and in part because of the limited field of vision of the tourists
>
> (2014: 411).

Thus the dominant media architecture of surfing spaces continues to script the fantasy of empty waves and absent cultures. Carlsen exemplifies this point through the case of the 2016 Brazilian surf movie *Une Filme de Surfe*. In this film, Carlsen argues,

> the audience did not see the complications of travel to Indonesia, they did not hear Indonesians speaking on camera and they did not see any cultural marker that identified the island as Indonesian. In addition, the film effaced cultural markers that could have established the surfers' "Brazilianness". Propelled by the filmmakers, the surfers engaged in ...hijinks, that reinforce the fratriarchal aspect of their constructed identities
>
> (2019: 239)[2]

In this film, the environmental costs of long-haul, long-distance, jet-engine travel were not engaged with (yet for a general discussion on this, see Anderson, 2017, for example), the particular cultures of the host destination were deemed irrelevant, and the specific geo-cyborg relations of the surfers were silenced, re-routed to depict a mobile surfing lifestyle. As Carlsen argues, in this construction, the Brazilians' cultural identity [wa]s replaced with a "translocal" surfer identity: a composite of contemporary Californian, Hawai'ian and Australian surfer identities

(2019: 239).

In this film, but also in surf media more broadly, the (script of) the travelling surf-rider, defined by frat-house camaraderie, adolescent humour, film-watching, and magazine-reading is transposed on top of individuals abstracted from their home and host cultures and emplaced within an imagined surfing space. Taylor describes the effect of his script on his own experience, stating:

> I thought it was a waste to travel thousands of miles to sit in a swampy guest house playing video games and watching DVDs. But I couldn't deride the lifestyle either. I was... just like the rest of the crew: a beer swilling vagabond surfer who'd gotten used to taking this lifestyle around the world with him

(2005: 131).

Thus when mobile, the surf-travel script encourages surfers to become "trans-local" in their identity and practices, conforming to the western archetype (see Chapter Ten) of rootless individuals who have little to champion beyond their commitment to stoke. Due to the lack of "incoming" and "outgoing" interactions between local cultures and surf breaks (see Chapter Four), surfers are no longer relational geo-cyborgs but become reduced by the "trans-local" code into naïve and irresponsible "tourists" (see Sheller and Urry, 2004; Heimtun, 2007).

The code and its consequences for "dream" destinations

The code of the trans-local surfer has consequences for the littoral destinations visited, in terms of both its broader cultural geographies, as well as its surfing spaces. As Appadurai (1996) and Sassen (2006) identify, when any trans-local practice goes mobile it interpenetrates specific locales and has the capacity to actively displace local or national ways of life. These consequences draw attention to the fact that the "open frontier" that is sold by the surfing space script, is in practice nothing of the sort. As the waves celebrated by media architecture become magnets for entrepreneurs, developers, and businesses who seek to extract dollars from surfing tourists, the extent and nature of this development threatens existing, local cultures. There isn't space in this chapter to document the extensive effects of surf travel in destinations around the globe, but perhaps it is useful to include one example. Regions such as Kuta and Legian in Bali for example (the "Disneyland" scripted by Rae above) have since their "discovery" in the 1960 and 70s become notorious for their "rampant development, low brow nightlife and crass commercialism" (Lonely Planet, 2010: 269). Although offering a "tropical paradise" on the waves, these regions have become a prime example of "the destructive effects of tourism" on the shore (Lonely Planet, 2001: 234). Barilotti sums up the development of Kuta in the following way:

> Kuta Beach started out as a drowsy little fishing village in the 1930s, catering to a small number of vacationing European colonialists. Its surf potential was discovered by Australians in the mid-1960s. Since then, it has morphed into a fully-fledged surf ghetto on a par with Huntington Beach or the North Shore... In our blind zeal to set up insular surf enclaves, we parachute advanced technologies into third-world economies and set up brittle unsustainable infrastructures. The list of soiled third-world surf paradises... is long and growing
>
> (2002: 92).

Across the archipelagos of South East Asia, including islands such as Nias and Bali, surf development is often intense. Summarised by Barilotti, across the region, "4000 years of ancient animistic squat culture [has] now smacked straight into Western techheavy materialism" (2002: 92). This has led to the cultural influence of western society colonising the traditions of endogenous regional and local cultures, as Barilotti explains with respect to Nias:

> the effect of surf tourism on the Niah, a proud, warlike tribe once notorious for their headhunting and elaborate costumed rituals, has sped the erosion and disappearance of traditional ways. Twenty-five years of cashed-up westerners tramping through Lagundi village has seduced the local youth with lurid Baywatch fantasies of the North American high life
>
> (ibid: 93).

In Bali, as local journalist Eric reports, "for centuries, Balinese women have obediently carried the responsibilities assigned by tradition. Now, however, tradition is becoming increasingly compatible with modern life. Women are beginning to ask questions about their own destinies" (2011: 30). Whilst western ideas of feminism may be seen progressive when compared to Southeast Asian patriarchies, substituting long-held traditions for industrialised poverty or employment in hawking, prostitution, and casual labour, all in the service of surf tourism, appears to be a dubious advancement. Thus as surfers are seduced by the surf-travel script and enact in practice a trans-local surf-rider code, the effects on local areas are, at best, mixed. Whilst they may bring the tourist dollar through their spending habits, trans-local surf-riders also bring cultural shifts, environmental pollution, overdevelopment, and crime (see Buckley 2002). As Severson reflects,

> Was [this travelling surfer script] good for surfing, and the world? At the time it seemed perfect... [and] some surf centers have provided ways out

of poverty… [but] I'm sure that surfers upset some cultural tenets and economic balances

(2014: no page).

Culturalism and racism in the surfing spaces script

As we have seen, the translation of the surf travel adventure script into a trans-local code abstracts surfers from their geo-cyborg relations and re-routes them into a simplified imaginary of the destination littoral. Surf riders are now coded as intrepid explorers, discovering new frontiers and imposing their own interests in new places. This preoccupation with their own values, and their disinterest of its effects on host populations, can be seen as a form of "culturalism". Culturalism can be understood as a prejudicial view, often based on ignorance, which caricatures and dismisses the interests and lives of one cultural group on the grounds of their difference to another's. Carlsen identifies such culturalism in his analysis of *Une Filme de Surfe* (noted above), in his words:

> … [the film] reduces the complicated cultural interactions of all surf travel to the idea of Westerners finding clueless locals living without any knowledge of the value of their own waves or even of the modern world
>
> (2019: 241).

Rather than the travellers checking their own privilege and choosing to respect the positions of those they are visiting, in this case local populations are given value (or not) only through the frames of reference of the trans-local. Just as the trans-local surf-rider is encouraged to ignore their own privileged position in broader networks of power, they similarly dismiss local populations' under-privileged position. As a result, the impacts of the surf-riders' presence can be decontextualised, becoming neutral outcomes of a natural process, rather than the specific consequence of the values and choices of the dominant actors. The presence of culturalism thus creates a situation where unprepared destination littorals are,

> homogenized by and for foreign surfers, where local people are seen as [either] obstacles to riding waves, valued [only] as means of production (labor), or at the very best,… treated as cultures to be consumed when the surf as gone flat
>
> (Ruttenberg and Brosius, 2017: 111).

In one sense, the consequences of the trans-local surf-rider code imposes a version of "natural selection" (or "survival of the fittest") on surf-travel (see Darwin, 1859, 1996). The implicit rules of the game centre on the ability of local populations to defend themselves from the incursions of trans-local surf-rider, or succumb to their culture. Given that the littoral "frontiers" explored by trans-local surf-riders are often in the majority world, and the surf-riders themselves are often white and derive from the minority world, it can be argued that this code is not simply culturalist in nature, but also has a racial dimension too. In what may be termed by some as "less-developed" and "non-white" worlds, due to structural processes emanating from the era of empire (see Anderson, 2021) endogenous surf-cultures are often less organised that those travelling, and the local economy remains more susceptible to the pull of the tourist dollar. In these circumstances, the perpetuation of the frontier myth by surf media architecture maintains the illusion that destination littorals (and their "non-white" populations) are merely tabula rasa, existing to be civilised by the intrepid ("white") surf-explorer. On one level, the trans-local surf-rider code thus becomes part of "a colonizing process of Western socioeconomic and cultural domination" (Ruttenberg and Brosius, 2017: 111), whilst imposing the "myth" "that surfing was [and is] a white activity, denying or erasing the potential for … black [and other] surfer subjectivity" (Wheaton, 2017: 181).

Indeed, in Wheaton's own research, her respondents identify directly how such "erasures" and "denials" are not neutral or natural processes, but are thoroughly pernicious and political; as she explains:

> Some of the black surfers [I] interviewed said that the postcolonial mentality in films and media travelogues was one of the aspects of surfing culture that most angered them. [As one respondent states,] "Because what you're saying is, you're going into these third world countries, [and assuming that] we are still uncivilized. You came, [and] you're going to teach us how to surf!… It amazes me to think that we are so… and I'm going to say [we in the] United States because I don't see it [this attitude] so much in other places, as we feel like we have to find everything. Our white culture has to discover everything. You don't need to discover us we've already been here
>
> (2017: 181).

The script of adventure travel, and the code of the trans-local surf-rider is therefore not without its cultural and political consequences. Its choice to promote travel, in certain ways, to particular places, does nothing to challenge existing colonising and commercial processes that rely on culturalist and racist assumptions about the world. This script and its codes, and the imagineering process as whole, is therefore not an escape from the mainstream culture of America or the minority world more widely, but is directly built on its inequities and prejudices.

The "erased" and "denied" fight back

In some cases, local surf-riding cultures seek to challenge the code of the trans-local surf-rider. Reflecting the code of surf-rider provenance, and running counter to the disembedded and simplistic imagination of the surfing spaces script, some local populations position themselves directly within the context of colonisation and racism outlined above, and employ this to defend their own protectionist tactics (see also Chapter Thirteen). In Hawai'i, for example, codes of provenance are well-established to defend against trans-local surf-riders, and are argued to be a necessary response to the multiple forms of colonialism perpetuated by white settlers on the islands. In this light, local groups like Da Hui (formed in 1976 and now led by Kala Alexander) impose codes of surfer-provenance that realise mixophobic prejudice to contemporary non-local surfers on Oahu North Shore. These are framed as gestures which challenge the historical and contemporary discrimination they themselves suffer. As Smith describes it:

> On the North Shore no one will ever come to a [trans-local surf-riders'] assistance, partly because haoles are hated for stealing this land from the local Hawaiians. The indigenous here were treated with the same sort of disrespect dished out upon the indigenous on the mainland. There is a history of stolen land, stolen children, raped identity. The associated pain and anger is reflected back on the white skins. The haoles. And though there are not many with Hawaiian blood anymore, the nonwhites from Mexico, the Philippines, Tahiti, wherever carry the mantle and dish out retribution
>
> (2013: 75).

Conclusion

The normalisation of the surf travel adventure script and trans-local surf-rider code has thus contributed to the dominant assemblage of surfing spaces which are far from being open and free. By imagineering surfing spaces as open frontiers, a cultural geography has been established which is directly connected to culturalism, colonialism, and racism, and has produced an orthodoxy which assumes these spaces to be "exceedingly" (after Dominguez Andersen, 2015: 512), "over-whelmingly and... undeniably" (after Olive et al, 2019: 150) white. On those occasions when attention is called to the prejudice at the heart of the surfing space script, two key responses are common. The dominant surf media architecture responds by drawing on the motivating factors that bridge and bind the imagineering process (see Chapter Eleven). Initially, the relational sensibilities generated by surf-riding are used to excuse any negative effects produced by its codes, as Thomas (when editor of *Surfer* magazine) explains,

> ... If you're a surfer, the promise of empty, perfect waves has the potential to short-circuit even the most high-minded moral compass
>
> (2014: 20).

Through such a response, the power of the extraordinary sensations generated by surf-riding are elevated further (and thus perhaps become more attractive to some, and validate the actions of others), but also reduce the surf-rider to somebody who appears to be at the mercy of their impulses and thus rendered incapable of responsible actions towards their fellow humans. The second response resorts to the economic "bottom line", suggesting that although these surfing space codes may be politically indefensible, they are nevertheless profitable, and thus their existence is excused. This pragmatic view is well summarised by "Paul" (a Mentawai boat charter operator) when he states,

> Everyone chases th[e] dream... I can't imagine that the big rag trade boys are going to give up on that idea. They're the ones who have built that notion and have milked it to the max and made millions and millions of dollars from it. It's good business
>
> (cited in Ponting, 2009: 178).

Indeed the normalisation of the trans-local codes are such that for surf-riders to call out these orthodoxies as unethical, irresponsible, and irrelevant, is an act of "epistemic disobedience" (after McGloin, 2017: 210). Yet in light of the insights from Part II of this book, it is clear that if any actors fail to do so, be they academics, surf-riders, sponsors, surf-lifestyle entrepreneurs, media architecture or other support industries, their failure renders them complicit in the outcomes of this ongoing prejudice. Thus as Colas suggests,

> as the World shrinks and frontiers retreat in the face of a frenzied search for perfect, uncrowded surf, [we should] take a moment to reflect on what it means to be a global surf traveller
>
> (Colas, 2001: 4).

It is thus possible and necessary to reflect on the script of adventure travel and the trans-local surf-rider code in order to imagineer them afresh. As Hough-Snee and Eastman comment, "there is much practical value in moving surfing culture and sport – the spoils of two and a half centuries of cultural colonization – toward a recognition and correction of their violent pasts" (2017: 2), and as they go on to say, "the last decade has seen the rise of countless activist initiatives pushing to recast surfing – past and present – in restorative feminist, queer, ethnically and racially inclusive, and decolonial modes" (2017: 2).

From protectionist practices in Hawai'i, "eastern" archetypes in new surf media, organisations which re-claim a space for themselves within surf-riding (including for example *The Black Surfers' Collective*, no date; and *brown girl surf*, no date), to collectives which engage directly with host populations in non-culturalist ways (see for example, Britton, 2015; and Surf Resource Network, no date, see also Brody, 2015), the establishment of more sustainable, convivial global communities are being trialled as working alternatives to the dominant script. Such activities are vital for the future of surfing spaces, a subject we will focus on directly in the final chapter of this book.

Notes

1 "It was the day after Valentine's Day and I was spending some time back home in Canada doing some much needed editing work at my office... only to hear that familiar 'pop in my web browser. It was Facebook chat. The message was from a friend of mine in Morocco. The cryptic message read 'Thursday is going to be the day'.... I quickly opened the surf forecast model on my browser and animated the North Atlantic. A massive low pressure was headed on a collision course with Ireland... Seas upward of 50 feet and wave intervals in the 19 second range... It was too late for me to chase this swell there. The best option would be a thousand kilometres south, in Africa and I still had time to make it. Only catch – I had three hours to make the flight.

....As I pulled into the airport, I thought how our technology is changing... A few years ago it would have been impossible to book a ticket through an iPhone on the way to the airport, let alone answer emails and look at surf forecasts... As I parked my car in the long term parking lot and pulled out my bags... I said out loud to myself, 'Too easy' with a big smile on my face.

...The next few days felt like a Utopic Groundhog day where the same conditions would repeat themselves. Upon returning... back home in Canada... suddenly this all too familiar noise popped up on my computer. It was Facebook chat. It was my friend just up the shore. He informed me this fabled right point break was absolutely firing and doing an impersonation of Rincon sprinkled with icing sugar... I got in my car and headed for the coast...." (Ouhilal, 2011: 88/95/97)).

2 As Carlsen explains, "John Remy calls the phenomenon of young men forming a male society through hijinks and pranks a 'fratriarchy'. According to him, this group is dominated by an age-set of men who have yet to take on the responsibility of having a family and are primarily interested in fun [see Remy, 1990]" (2019: 246). As Carlsen goes on, "with some notable exceptions, women make infrequent appearances in extreme sport films. They are usually attractive and silent consorts of successful male athletes, implying that these young men are irresistible to women, but prefer to spend most of their time without them" (2019: 248).

17 New Surfing Spaces

Introduction

As we have seen in this book, surfing spaces are worlds of practice which involve highly skilled humans riding waves of energy as they pass through water. However, this book has argued that the act of surfing does not exist in isolation, it is defined in part by the cultures that associate themselves with it – by the individuals, ideas, myths, writings, images, media, and merchandise which at once reflect, influence, create, and commodify the practice. The book has therefore approached the practice of surf-riding and its dominant cultures as mutually definitive; surf-riding is not only a practice, but also a world of power relations, inequalities, conflicts, and geographies.

From this basis, the book argues that the emergence of surfing spaces are not natural processes but cultural ones. Surfing spaces are composed through ongoing processes of cultural ordering and geographical bordering, where one cultural group's ideas of "good" and "bad" surf-craft and of "appropriate" and "inappropriate" surf-riders are transformed through practice from being merely one alternative, into the only way to be. Surfing spaces are therefore the ongoing composition of different ideas of who surf-riding is for, what it should be, and how it should be understood.

This book has made its own contribution to this ongoing composition. In Part I, *Assembling Surfing Spaces*, this book critically examined the ways in which the physical geographies of surf-riding assemble with human practices in order to create surfing spaces. It introduced the littoral zone as the primary space of surf-riding, framing it as a space of energy flows which come to co-constitute many humans' sense of self. This identification is further enabled by the different surf-riding technologies employed by humans, and through their assembled coming togethers, co-produce a range of relational sensibilities which are popularly re-presented as "stoke". Part I argued that the co-constitution of these physical geographies and human practices should be considered as eventful in nature, transforming the ways in which surf-riders, but also perhaps the audience of this book, understand and orient themselves in relation to the world.

In Part II, *Scripting Surfing Spaces*, the book examined the ways in which the transformative encounters of surf-riding have been (b)ordered by those

DOI: 10.4324/9781315725673-17

involved. This book introduced the notion of "imagineering" to refer to the process of imposing order on and giving meaning to the eventful nature of surf-riding with the view to control and commodify its associated cultural geographies. This imagineering process wove the relational sensibilities of surfing spaces into a broader "script" which strongly resonated with popular western archetypes, and harnessed media "architecture" to disseminate it to key surfing markets around the world. Influenced by imagineers and their burgeoning constituencies, the "surfing spaces script" in turn created its own "cultural codes" for surfing spaces. These codes of positioning and positionality imposed specific (b)orders on surfing spaces, coming to define the lived experience of this activity in practice.

As we have seen in this book, this imagineering process was immensely successful, and from its inception in the 1950s through to the present day, its script and codes gained then retained a hegemonic influence over the nature of surfing spaces. In this final chapter, we take stock of the influence of the script and its codes in the contemporary context. We suggest that although the surfing spaces script remains intoxicating for many, the cultural, political, and media context in which it circulates has significantly changed, and the contradictions in its (b)orders have become increasingly difficult to sustain. The chapter suggests, therefore, that in the third decade of the twenty-first century, the hegemony of the surfing spaces script is weakening. It makes this case by focusing on two interrelated processes: firstly, the diversification of socio-demographics and craft which now actively practice surf-riding, and secondly, the new media architecture which is used to control and communicate these activities. These two interrelated processes offer an economic and existential threat to the dominant imagineering process, whilst raising questions about what the future of surfing spaces may be.

As explored in Chapter One, new surfing spaces are continually being formed through natural and human-induced processes in littoral locations, whilst new populations are switching on to the pleasures of surf-riding in different regions of the world. As the waves occurring in different stages in the hydrological cycle are being assembled for surf-riding (for example, lakes, bores, and river waves), surf-entrepreneurs are also seeking to engineer waves in purpose-built lakes and pools, creating new opportunities to privatise the experience based on the ability to pay. With these new surfing spaces being enabled by the success of the script and its constituent codes, what will the future assembling and imagineering of surfing spaces look like? What will the surfing space script and its codes evolve into in the future? As this chapter engages with these questions, it begins by reflecting on the dominant codes and script that have been examined in this book. It suggests we can conceptualise the dominant paradigm of surf-riding to date as a *"stem"* script, developing from the *"roots"* of endogenous histories and the commodifications of Hawai'ian surf-riding by nations and individuals, and *"branching out"* into new archetypes and actions in recent years.

The "stem" script of surfing spaces

As we have seen in Part I, this book has suggested that all surfing spaces are assemblages of a range of processes and forces-in-elements. Surfing spaces are coming togethers of not simply materialities but of energies, they are worlds of transfer and flow, transiency and ephemera. So whilst it may be tempting to consider surfing spaces as fixed and immutable "things", they are better understood as spaces of "precarious relational accomplishment" (Philo, 2005: 824).

This understanding applies not simply to the physical composition of surfing spaces, but also their cultural counterpart. In practice, this means that what surfing spaces are depends on how they are harnessed, actioned, and maintained; in other words, how they are "performed". Thus whatever surfing spaces are at any point in time is not natural or neutral in nature, but always up for grabs and open to new performances; in short, surfing spaces are, "all subject to, and productive of, the influence of power" (Cresswell, 2019: 197). Whichever cultural group is successful in wielding their power in and through surfing spaces gets to take and make its cultural geographies, influencing how we consider, "what is right and wrong, good and bad, normal and abnormal, beautiful and ugly" (Jordan and Weedon, 1995: 5) in the surf-riding world.

In this book we have seen how the physical geographies of surf-riding and the relational sensibilities associated with them are eventful for human beings. We have also seen how this potent experience has been taken and made by particular cultural groups through a process of imagineering. This "imagineering" was undertaken by a small group of young, white, males in California in the 1950s and sought to impose order on and give meaning to the cultural geographies of surf-riding. In this final chapter we suggest we can understand this process as producing the "stem" script of surfing spaces.

In this context, the word "stem" should be considered as both a noun and a verb. As a noun it refers to the "main body" of a thing or set of ideas, in this case, the main spaces of surf-riding as scripted and coded by the imagineering process set in motion in the 1950s. This "stem" script grew out of its own particular "roots" of surf-riding: the colonisation of endogenous surf-riding by British then American nations in Hawai'i, the commodification of this reduced and recuperated culture by capitalists such as Hume (see Chapter Nine), and the creative engagement with these iterations by individuals such as Blake (ibid). This "main body" then came to dominate the spaces of surf-riding, silencing and marginalising competing "bodies" (i.e. other endogenous surf-riding histories and practices) through the power and persuasiveness of their own values and practices. As surf-historian and -journalist Barilotti puts it, it transformed surf-riding from a "cult" to "culture" (after Barilotti, 2013: 15), influencing,

> surf language, surf music, surf art, surf media, surf fashion – all the basic elements of what are now considered essential to modern surf culture were …codified within this brief window of time
>
> (Barilotti, 2013:16).

The "stem" script of surf-riding thus came to be define the main body of surfing spaces in the remainder of the twentieth century. In detail, the "stem" script – or perhaps the acronym "STEM" in the sense it will be used here, can be understood to be constituted by the following components. Initially, by "Stoke". Due to the imagineers of the stem script understanding the relational sensibilities of surf-riding through their own involvement in the practice, they knew that the energies-transferring through littoral locations offered an opportunity for humans to feel alive, and this could be harnessed into a broader framework of meaning for the practice. The imagineering process thus drew attention to the eventful encounters of surf-riding and connected them to broader values which resonated with western cultures. Ideas of fun, freedom, and exploration assembled with notions of entrepreneurship, elitism, and prejudice to compose a script and impose codes from which "stemmed" the hegemonic definition of surfing spaces until the present day. A key element of the "STEM" script was its illusory nature. Rather than embedding surf-riding within the realities of contemporary economic and political life, the "STEM" script "disembedded" (after Ponting, 2009) the practice of surf-riding from histories of colonisation and its defining contexts of ablism, racism, and patriarchy. The result was a script of surf-riding that ignored inequalities and in so doing attempted to neutralise them, presenting a world of unproblematic open frontiers which (certain) surf-riders were free to explore without concern for the effects on existing populations, marginalised others, or the broader (cultural) environment. A second key component of the "STEM" script was therefore the cultivation of the "Trans-local" surf-rider, a cipher who was deemed entitled to externalise all inconvenient networks and relations in order to personally realise the stoke inherent to the "STEM" script. This imagineering of the "STEM" script was motivated by its third key constituent part, "Entrepreneurship". The script was created by a group of white, young, male Californians – a "small bro-ish" cadre in the words of Woody (2015: 71) – in an attempt to monetise surf-riding and preclude their need to gain conventional employment. Through their imagineering they invented a new role of the "surf-entrepreneur" who bridged the poles between profession and pastime in order to create a thresholder role for themselves both within and beyond the culture. Through prioritising their own goals they instigated a "gold rush" to surf, encouraging others to not only get involved in surfing spaces, but also further commoditise them, regardless of the broader consequences. This approach to surfing spaces can be seen in the final constituent part of the "STEM" script: "mixophobia". The "STEM" script tended to valorise particular craft and bodies in surfing spaces (from the Short-board, to Trans-Local surf-riders, to Ethnic elites, and Male surf-riders), generally coding them as a monoculture which traded in loyalty to its core group and actively discouraged, marginalised, or silenced (i.e. "stemmed") those who did not fit their narrow codes of authenticity. Due to the dissemination and success of the STEM script, the majority of surfing spaces were composed in its image, and against which more marginal and other surfing spaces can be defined (both historically, in the present, and in the future).

The "stem" (or STEM) script has had clear consequences for surfing spaces. It normalised particular idea(l)s of surf-riding that have, over time, engrained themselves so deeply "in many surfer's DNA that… they are not even conscious of them anymore" (after Pepin Silva, see Chapter Twelve). To paraphrase Jordan and Wheedon, it is due to this script that,

> we come to accept that men are better [surfers] than women, that Black people are less [able to surf-ride] than Whites, that rich people are [entitled to the best surf] because they have worked hard and the poor are [less entitled] because they are lazy
>
> (1995: 5).

But as we know, it is always possible for (b)orders to be crossed, values to be challenged, and new cultural geographies to emerge. As scholars and surf-riders it is perhaps essential for us to question authority and bring about "an[y] opportunity for thinking differently about something we might otherwise take for granted" (Sargisson, 2000: 104). Such critical reflection opens surfing (and all) spaces to alternative futures, enabling the possibility of scripting and coding them differently. The chapter continues by identifying two interrelated processes which together have challenged the hegemony of the "stem" script of surf-riding; the diversification in types of people surf-riding, and the new media architecture in which their idea(l)s are communicated.

Contemporary challenges to the "stem" of surf-riding

In the mid-twentieth century, the "stem" script presented surf-riding as an escapist fantasy by young white men, for young white men. Yet despite the exclusionary nature of this script, becoming part of the surfing space assemblage "retains a primal hum…" for a much larger populace, with its lure remaining "beautiful, if not unaltered" by the imagineering process (Hulet, 2011: no page). The growth in sheer numbers surf-riding, driven in part by diversification in the types of craft adopted meant that although the "stem" script and its riders dominate in many places, their hegemony was now open to question. As Woody sums up:

> Paddle out into today's lineup in multicultural California, Hawaii, Australia and many other places […], and you'll encounter a far more diverse wave tribe. Sure, the sun-bleached blonde, adrenaline-addled descendants of Miki Dora[1] are still there ·in force. But there are far more women, 40-something groms, gay surfers and people of all races and ethnicities as well as parents surfing with their kids. There's the gray-haired grandmother riding a longboard…, the teenager… surfing a homemade alaia… and the middle-aged guy on a US$1,300 Danny Hess salvaged wood and-recycled-EPS-foam board…
>
> (Woody, 2015: 70/71)

Thus despite surfing spaces being scripted for the young, white, male[2], the allure of the relational sensibilities involved in surf-riding remain attractive to many; as Bush identifies,

> Though specific numbers are hard to come by[...], industry sources estimate that 60 percent of surfers in the United States are over the age of twenty-five and that women are now entering the sport at a significant rate, comprising as much as one-third of the estimated 2.3 million surfers in the United States (Board-Trac, 2009; Higgins 2007; Lewis 2010; Wagner, Nelson and Walker 2011; Warshaw 2008)
>
> (2016: 291)[3]

Thus the world in which the "stem" script existed was changing around it. The socio-demographics who wished to participate in surfing had diversified, and this change was compounded by the way these participants consumed their media architecture.

As we have seen in Chapter Eleven, there have been changes within the traditional media architecture of surfing spaces. New scripts have emerged to cater for different constituencies, offering alternative meanings and values about what surf-riding is and what the authentic surf-rider can be. More "eastern" in its archetype, more inclusive to different crafts, more collective in its orientations, and more communal in its sense of stoke, this script welcomes difference in the line-up, as well as revealing the dominant script to be far from natural and neutral in its imagineering, but rather as a framework which "stems" the development of surfing spaces themselves. This recognition, occurring hand-in-hand with different (or maturing) socio-demographics in surf-riding, has had the effect of reducing the monopoly of message disseminated by the dominant script-writers, and liberated their constituencies to look for meaning in their practice elsewhere. This diversification has drawn attention to the "inertia" of the dominant script, highlighting the "hideous disconnect between the way women are represented in the [dominant] surf media, and the actual women who actually surf" (Davies, no date), as well as the broader variety of surf riders that "you won't see reflected in the [dominant] media" (Woody, 2015: 71).

> ...print editions are in decline, and the digital business model is taking over
>
> (*Surfer Today*, no date).

The emergence of new variations within conventional media architecture has occurred concurrently with developments in new media technologies. Media architecture which, until a short time ago, used magazines to "reach wide audiences" across the surfing world (after *Surfer Today*, 2013), are now being challenged by the rise in digital communications. These technologies pose a threat to the business model of traditional media. With apps and social platforms, enabled by personal and purchase algorithms, individuals can have tailored content pushed to their cells more cheaply, regularly, and conveniently, simply via a swipe of their screen. In this context, individuals now have choice. Why might they continue to subscribe to periodical magazines that no longer represent their experience of surfing spaces, when you can get daily free media that does?

The rise in recent decades of social media platforms has thus substantially challenged the means, message, immediacy, and cost of media architecture for the surf-rider. This new media has the potential to diversify the dominant media's script and challenge its monopoly. No longer is the power to control the script and its codes centralised within a small cadre of influencers (including editors, photographers, and advertisers); now every surf-rider has the capacity to communicate their practices around the world, developing the flow of information from a uni-directional, hierarchical, and arboreal structure, to a rhizomatic, many to many, plurality. This challenge to the dominant architecture of surf media has been swift, but may be significant. Replicating the "brief window of time" in which the dominant culture of surfing was established in the 1950s (see Barilotti, above), since midway through the second decade of the twenty-first century (when Jean-Sebastien Estienne, European Marketing Manager of RipCurl, stated that, "we are deeply convinced that surf magazines are amongst the best tools we can use to reach our core customers" (Research Interview, 2015)), many keystones in the architecture of surf media have ceased to trade (including *Surfing* magazine which closed in 2017 and *Surfer* in 2020). Facing falls in subscription, an over-reliance on advertising, and an inability to capture the changing nature of the surf-riding line-up, these closures are seen to herald, "the end of surfing's golden era" (*Surfer Today*, 2020); with their demise comes the acknowledgement of how fundamental these architecture were in imagineering the popular value and meaning of surfing spaces, as *Surfer Today* notes, the end of these magazines signals, "the death of a vital component of surfing's original essence and culture" (ibid.)[4].

With the advent of new media, surf-riders are no longer reliant on traditional gate-keepers to access professionals, photographers, products or opinions; if we want more information about "the source" by the shore (after Taylor, 2007a and b), we can go straight to the source on social media. Cultural influencing has been democratised, with merchandisers, athletes, local surf-riders, or photographers able to post their own views directly to their audience. Where once it was "John Severson's Surf" (Severson, 2014, see Chapter Eleven) which scripted surfing spaces, now it can be yours, or theirs, or individuals like Chris Burkard's (https://www.instagram.com/chrisburkard/). The power, the influence, and the income, is now taken and made by those who know how digital media connects to its audience, as Jeff Divine states,

> who are the most famous surf photographers now? Well, it might be Zak Noyle, he has like 400,000 followers on Instagram. That dwarfed a magazine circulation like *Surfer* and he's putting that stuff out all week long and he's linking product into it and making money off it… I mean

really the most famous surf photographer in the world is probably Chris Burkard... He's kind of taken the ball and ran with it... he takes a great shot then posts up that it was shot with these new attachments

(Research Interview).

In practice, therefore, the advent of social media fundamentally changes who can influence the nature of surfing spaces, as Divine confirms:

... [in] the old days... it was just a handful of people that created the culture. Everything they shot became the reality of surfing. [But now there is social media] so there's no more singular guy that is creating the reality. The reality is now created through everybody

(Research Interview).

In this new context, vital questions for the future of surfing spaces emerge. How will these new media be used to influence the culture and geographies of surfing spaces? Who will emerge as key powers? What new scripts will be created, and what new codes? How will surfing spaces be re-constructed, re-imagineered, and re-produced, and what new elites will be formed? Early research in this area suggests, for example, that diversity may be the hallmark of any new surfing spaces script, as Olive reports:

The social media app Instagram has become a popular everyday way to share visual representations of surfing culture and experiences. Providing an alternative to mainstream surf media, images posted on Instagram by women who surf recreationally both disrupt and reinforce the existing sexualisation and differentiation of women in surf culture

(2015: 99).[5]

Indeed, these new media, along with the new demographics involved in surf-riding, suggest that existing codes which limit and reduce "proper" surf-riding craft no longer reflect the more "fluid" geo-cyborg relations that many surf-riders identify with. New surfing spaces are, as a result, increasingly "polymorphous" (Stranger, 2011: 109), as new surf-riders shed allegiances to scripts (be they western or eastern) that ascribe craft with gender, acknowledging that no code can fully capture the diversity of surf-riders, or perhaps even the diversity of thoughts, philosophies, and urges within the plurality of the individual geo-cyborg. As a consequence, craft options (see Chapter Fourteen) increasingly co-exist as "gender-neutral choices within the subculture's range of styles" (Stranger, 2011: 107).

Diversity continues to proliferate with new technologies too (hydrofoils anyone?), as well as ways of combining craft to access surf in different areas of the ocean (for example "tow-in teams" working to surf new waves in more off-shore locations). Such diversity raises questions in relation to how new values and meanings may be associated with these practices, and how balances may be created between performances of conviviality and existing codes of positioning and provenance. Such issues of balance also emanate with respect to the new types of "fidelity" suggested by these new performances (see Chapter Eight). The "fidelity" of the surf-rider to surfing spaces remains profound – littoral-humans feel compelled to experience again and again the relational sensibilities produced through being part of surfing spaces, but as surf-riders age and their broader responsibilities become harder to ignore, the nature of this fidelity is also likely to change, as Ellis suggests,

> We're all older now, settled, families, divorces, kids, all the things life brings, but there's that one constant: we are surfers. ...saltwater lifers... marching to the sea to enjoy and appreciate every long paddle, ice cream headache, sunburn, and lipsmack we can cram into already busy lives
>
> (Ellis, 2011: 130).

In this context, new questions emerge in relation to how people orient their lives around the assembly with surfing spaces, and how these new strategies of fidelity fit with existing, more obsessional and absolute modes (see for example Bush, 2016[6])? There is considerable potential for scholars to explore vital new questions about how lives are co-constituted, compromised, and balanced by littoral encounters, as Pepin Silva confirms:

> I get why professional surfers surf, they're paid for it, that's their job. But why some mom with two kids who carves out one hour a week to get in the water to me that special, that takes determination, and that to me is more interesting
>
> (Research Interview)[7]

Branching out: New scripts in new spaces?

Perhaps one of the most crucial set of questions with respect to the new scripts and codes of surfing spaces is the influence of different physical components in the surfing space assemblage. As surf-riding expands in new regions of the world, how will conventional littoral locations be taken and made by new cultures (especially for example in large population bases like India, or China (see Evers, 2017)? What will the influence of the "stem" script be in these locations, how will it be indigenised or re-imagineered in these new cultural contexts, and what reciprocal consequences may be created for the "stem" script as a whole? As noted in Chapter One, different (non-saline) stages of the hydrological cycle are also now surfed by a variety of craft. How might surf-riding on rivers, bores, and lakes be mainstreamed or marginalised in the future, and what codes of performance may be (b)ordered? Despite the importance of these questions, as Chapter One suggested, perhaps the biggest challenge to existing performances and scripts is the advent of artificial waves.

> [Artificial waves are] the biggest invention in surfing since the surfboard
>
> (Haro, 2015: no page).

Humans have long made waves, either by accident or design[8]. However, given the complexity of wave generation, it has proven an elusive science in the littoral zone[9]. One way to control the variables at play is make outdoor "laboratories" in which waves can occur, harnessing advances in technology and engineering to manufacture waves at the touch of a button. Indeed, the advent of artificially generated waves in "pools", "gardens", and "ranches" is increasingly widespread. These surfing spaces are not anchored to littoral locations and can be situated anywhere with the correct geological characteristics. Often sited inland, they bring the opportunity to surf-ride to many land-locked populations. However, what are the consequences of these new surfing spaces for how we understand the nature of the culture and its practice?

Artificial surf-lakes create controllable, predictable waves. For the price of admission, a surf-rider is granted the right to catch these waves without competing in a line-up, or having to develop any vernacular knowledge to be positioned correctly; as surf-rider and marine biologist Scales states at a wavegarden in Spain,

> ...For years I've struggled to be in the right place at just the right time to catch an awesome wave like this. After a few more goes I begin to get the hang of these artificial waves which sweep up and down the lagoon, one every minute
>
> (2015: no page).

As Scales suggests, artificial waves mean that the surf-rider no longer needs patience to wait for waves to come, or the knowledge to read weather maps, understand swell conditions, or the idiosyncratic coming together of land, sea, and air in order to catch waves; in "laboratory" conditions, if you miss a wave, you can just buy another (or another session). Artificial waves therefore change the assemblage of physical components that constitute surfing spaces, and the relations that humans have to them. Artificial waves are generated in (relatively) safe, sanitised environments, without other surf-users, non-human bodies, pollutants, or natural processes (like rip-tides) negatively affecting the experience. These surfing spaces may therefore lack the sublime elements of the littoral or their risks. In this context, different craft may be encouraged, with the popularity of shortboard moves consolidated, as surf-journalist Haro suggests,

> If surfing is going to maintain its physics-busting aerial progression, wave parks are likely to be where it'll happen. Blow out of a perfect liquid ramp at J-Bay just to try a backflip or something mad you know you won't stick? Nah. In a wave pool, when you know the next wave will line up identically... Well, why not?
>
> (2015: no page).

In this context, waves now have a direct market price (rather than simply a cultural value), and positioning can be bought rather than earned. The breadth

216 *New Surfing Spaces*

of knowledge required to successfully surf-ride is concomitantly narrowed, but the opportunity to learn accelerated. The predictability of wave pools therefore may create the conditions that surf-riding as "acrobatics" may flourish, in turn enabling a more consistent and open competitive environment, where moves can be judged, and broadcasters (and their armchair spectators) enjoy drone-eye views of the athletes at play. Moving waves from the littoral to the lab, surfing through science if you will, thus offers a fundamental challenge to the person–place relations, the relational sensibilities (where functional flow may trump the surfing sublime), and the conventional geo-cyborg ontologies of surfing spaces. With artificial waves, there is no waiting, no wild, just waves.

It is possible to conjecture that artificial waves may also (further) out-mode the stem script. As stated, enjoying waves "to order" does not require any ability to explore, simply the ability to pay. The replacement of the "rugged individualist" with the "restaurant surfer" (Haro, 2015) who orders up their wave and has it delivered in an instant is thus a significant challenge to the "stem" script. Such "fast-surfing" (following the trend in fast-food) could be seen to offer a more open and democratic future for surfing spaces, resonating strongly with the arguments mobilised to defend the privatisation of breaks more broadly; as professional surf-rider Shane Dorian states (in relation to a privatised break in Oceania), surfing is now:

Figure 17.1 and 17.2 Artificial Waves in Wales (Images: Lyndsey Stoodley).

open to anyone [who pays] ... when you drop [...] your money [...] you're guaranteed your chance to catch an epic wave

(in Jarvis, 2014: 74).

Yet such arguments can also be seen as an attempt to naturalise and neutralise the effect of market forces, and ultimately perpetuate an unequal (surf-riding) world. As we have seen, access to resources, opportunity, and income is often not a meritocracy, but a consequence of historical processes of othering, colonisation, and prejudice. The imposition of the market in these circumstances will intentionally and inevitably benefit those who are already at an advantage in relation to these structures. In this context, questions arise as to what extent these new surfing spaces will become playgrounds for middle class elites, and what effects that may have on the future of littoral locations[10]. In this context, how will the "stem" imagineers respond to the rise of artificial waves, given the challenge they pose to the adventurer archetype, but the potential they offer for new sources of income?[11]

Conclusion

As this book has demonstrated, the practice of surf-riding is enduring and compelling. The allure and magnetism of surfing spaces continues to attract participants whilst remaining resilient and "sturdy enough to survive mercantile attempts to festoon them with every gaudy, five-and-dime gimcrack under the sun" (Divine, 2011: 2). But in this changing context of increased participation, diversity in socio-demographics, pluralisation of media, and proliferation of surf-riding locations, what is the future for surfing spaces?

In the ongoing performance of surfing spaces, where the dominant "stem" script is being challenged, and new craft and cultures are emerging, it has never been more important for all surf-riders (be they active or armchair) to ask themselves two vital questions; in the words of UK surf-rider and -shaper Steve Croft:

"Why do I surf?" and – more importantly – "What is good surfing?"

(2008: 98).

Will the future of surf-riding be solely associated with fratriarchies, where selfish, aggressive males maintain their position through threat and intimidation? Should codes of fidelity that valorise obsession be rendered obsolete for contemporary life, and could more surf-riders (in littoral locations or on artificial waves) acknowledge their broader privilege within wider networks of (in)justice? Whilst reflecting on these vital questions, I stand in the sea, with the salt-water lapping my legs. I am reminded of my co-constitution with surfing spaces, the connections they offer me to the past, but also the opportunities they afford me for my present, for the chance of re-creation, of not only myself, but my position in the world, and perhaps even the future of the world itself.

Notes

1 According to Finnegan, Miki Dora was "the undisputed king" of Malibu, "a darkly handsome, scowling misanthrope with a subtle style perfectly suited to the wave" (2015: 88). Dora was a good example of the tensions implicit within the "stem" script, benefitting financially from the popularisation of surfing whilst protecting his own interests, almost at any cost. As Finnegan states, on the waves, "he ran over people who got in his way and scorned the mindless surfing masses in well-turned quotes in the mags, all while flogging [them] his signature-model surfboard, Da Cat, in adjacent ads" (2015: 88). For the surf media in general, Dora was the stem-script's (anti-)hero (see *Drift*, no date), inconsistent in his politics, but right-wing in his leanings. As Laderman argues, Dora "was a notorious bigot with a soft spot for fascism" (2014: 49, and following Westwick and Neushel, 2013). As Laderman explains,

"Dora fled the United States in the 1970s, spending much of his exile in apartheid South Africa, where, writing from 'deepest, darkest Africa' in 1986, he intoned that black South Africans are 'not like the blacks in the US who just kick your ass and take your wallet. These MFs are flesh-eaters. Give these guys the rights and you'll get white-man jerky for export'" (2014: 49).

In his politics Dora echoed the racisms of *The Endless Summer* (see Chapter Eleven), and remains an oft-referenced and celebrated figurehead of the stem-script (see, for example, Garcia, 2018).

2 Surfing remains scripted for the young, as summed up by this interchange in the 1991 film Point Break,

"At the other end of the counter, 15 is ringing up Utah's board.

15"Hey, man, guys your age learning to surf, it's cool, there's nothing wrong with it"

UTAH: "I'm twenty-five."

15"See that's what I'm saying, it's never too late."

(Point Break, 1991, Director: Kathryn Bigelow).

3 Market research also suggests that surf-riding constituencies are aging. In the noughties in the UK, the "core" magazine subscriber was between 35–44 years old (*Drift Magazine*, Research Interview), and as Woody records, "at one of California's more iconic breaks, Trestles, the median age of a surfer is 37" (2015: 71).

4 Indeed, in interview, Drew Kampion reflected on the need for the (once) dominant surf media to acknowledge its role in producing what this chapter has called the "stem" script:

"One of my realisations this year, which is way overdue, I think I felt it, I just never formulated it clearly, that capitalism without conscience equals cancer. If you have capitalism but you don't have value, conscience being the picture of aesthetic evaluation and emotional evaluation, you know, real values, if they are not connected with the capitalistic engine, then it's just a cancer it going to eat everything because it's a bottom line thing – we still use everything up" (Research Interview).

In the same vein, Tetsuhiko Endo, the surf editor of The Inertia (an online lifestyle sport zine), reflects on the risk involved in imagineering:

"Surf magazines are, at their best, caretakers of wave riding culture, acolytes of the ridiculous, but wonderful notion that we should take our recreation seriously... [but which can] quickly sicken and wither into a bloated, shambling corpse of corporate marketing campaigns and industry propaganda" (2013: no page).

5 The impact of social media may also have the effect of reducing the broader experience of surf-riding into "instagrammable" moments (reflecting processes in hyperreal worlds more broadly, see Roberts and Ponting, 2018, and Molz and Paris, 2013). As Kyle Thiermann (professional Californian surf-rider and podcast host) states when reflecting on individuals who may drop in on other surf-riders, "even if the guy (or woman) who drops in doesn't mess up the surfer who is deeper, they

are still going to [spoil] the photo of what could potentially be the biggest wave of their life, and that's pretty lame" (in Rode, 2020: no page). In an instagrammable world, it appears that relational sensibilities take second place to how the world is framed, captioned, and disseminated to followers.

6 As Bush records,

"The women in my study went to great lengths to negotiate and meld multiple identities (spouse, partner, business owner, employee, mother, friend) while being protective of the central role surfing plays in their lives" (2016: 300).

As she continues:

"Women with young children or pets will often negotiate childcare or petcare with their partners in order to surf. One woman I met negotiates nights of the week when her husband can play tennis in exchange for her surfing on other nights while he stays home with the kids. Women also use babysitters and nannies to allow them to surf. …women used surfing to bond with their children and families. [As one respondent states:] '… instead of being home watching TV, we're out sitting in the middle of the ocean seeing dolphins go by or just doing something that when you take off you're just like YES!'" (ibid.)

7 This would continue the work by Dolnicar and Fluker (2003), Sotomayor and Barbieri (2016: 63), and Rinehart (2015), as well as echo market research which suggests that even the "stem" surf-rider now only surfs a couple of times a month, "whenever he [sic] can find the time in between juggling a young family and work" and his interests in "skiing and mountain biking" (*Drift*, Research Interview).

8 Tom Anderson narrates how Napoleon helped create the surf in Hossegor, France through planting forest on coastal marshland (2007: 60), whilst the surf off the Gold Coast, Australia is formed in part through managed sand dredging offshore (see Stoodley, 2020).

9 As Scales reports, of the 12 artificial surf reefs built worldwide, none have worked. According to oceanographer Mark Davidson, "The flaws could come down partly to the turbulent sea shifting the [submerged infrastructure] until they lose their original shape. It's also clear that scientists still haven't yet got to grips with the complex mix of ingredients that go into sculpting natural waves" (2015: no page).

10 Surf journalist Haro suggests that surf-riders who prefer littoral locations should welcome the advent of the "restaurant" surfer (see below); in his words, "…wave pools far from the ocean, with their perfectly controllable, judging-friendly, sport-over-soul surfing mentality will clear […] your lineups, putting the plug in that influx of middle-Americans flocking to the coast, and creating a whole new surfing dynamic – one that keeps the soul in the ocean and the sport in the wavepools" (2015).

11 The "stem" surf media appear to making their peace with artificial waves, at once acknowledging their inevitability (as surf journalist Jarvis rhetorically asks, "in the increasingly crowded surf world is pay to play inevitable?" (2014: 72), whilst accepting that any allusion to a *non-entrepreneurial* surf culture is as illusory as the script itself; as surf journalist Clark states, "Many of us like to think of ourselves as belonging to some sort of ocean-worshipping tribe, dedicated to the pursuit of chasing swells and riding waves…. Seriously? Surfing sold out and went commercial a long time ago" (Clark, 2014).

Bibliography

Adey, P. 2010 *Aerial Life* Routledge: London.
Agnew, J. 1995 The hidden geographies of social science and the myth of the "geographical turn". *Environment and Planning D: Society and Space* 13 379–380.
Aguerre, F. 2015 Surfing, sustainability, and the pursuit of happiness. In Borne, G., Ponting, J. eds *Sustainable Stoke. Transitions to Sustainability in the Surfing World*. University of Plymouth Press: Plymouth. 34–43.
Albanese, C. 2002 *Reconsidering Nature Religion*. Trinity Press International: Harrison, PA.
Alderman, D. 2002 Writing on the Graceland wall: On the importance of authorship in pilgrimage landscapes *Tourism Recreation Research* 27 27–33.
Alderson, A. 2008 *Surf UK. The Definitive Guide to Surfing in Britain*. 3rd edn Wiley Nautical: Chichester.
Allen, M. 2007 *Tao of Surfing*. iUniverse: Lincoln.
Anderson, A. 2011 *Surfing Adventures. 100 Extraordinary Experiences in the Waves*. Wiley Nautical: Chichester.
Anderson, B. 2018 Cultural Geography II: The force of representations. *Progress in Human Geography* 43 6 1120–1132.
Anderson, B., McFarlane, C. 2011 *Assemblage and geography Area* 43 124–127.
Anderson, J. 2002 Researching environmental resistance: Working through Secondspace and Thirdspace approaches. *Qualitative Research* 2 3 301–321.
Anderson, J. 2004 The ties that bind? Self- and place-identity in environmental direct action. *Ethics, Place and Environment* 7 1–2 45–58.
Anderson, J. 2009 Transient convergence and relational sensibility: Beyond the modern constitution of nature. *Emotion, Space, and Society* 2 120–127.
Anderson, J. 2012 Relational places: The surfed wave as assemblage and convergence. *Environment & Planning D: Society & Space* 30 4 570–587.
Anderson, J. 2013 Cathedrals of the surf zone: Regulating access to a space of spirituality, *Social & Cultural Geography* 14 8 954–972.
Anderson, J. 2013c *Surf C* YouTube http://youtu.be/MaJpPSwNNF0.
Anderson, J. 2013d *Surf D* YouTube http://youtu.be/YLZTPK4XBTg.
Anderson, J. 2015 Mapping the consequences of mobility: reclaiming jet lag as the state of travel disorientation. *Mobilities*. 10 1 1–16.
Anderson, J. 2017 Surfing. A ritual with consequences. In Borne, G., Ponting, J. eds *Surfing and Sustainability*. Routledge: London. 172–198.
Anderson, J. 2021 *Understanding Cultural Geography: Places & Traces*. Routledge: London. 3rd edn.

Anderson, J. Erskine, K. 2014 Tropophilia: A study of people, place and lifestyle travel. *Mobilities* 91 130–145.

Anderson, J., Peters, K. eds 2014 *Water Worlds: Human Geographies of the Ocean.* Ashgate: Farnham.

Anderson J., Stoodley, L. 2018 Creative compulsions: performing surfing as art. In Roberts, L., Jones, K. eds *Water, Meaning, and Creativity.* Routledge: London. 92–113.

Anderson, J., Stoodley, L. 2019 The call of the sea: How sound co-composes the place of the surfed wave. In Doughty, K., Duffy, M., Harada, T. eds *Sounding Places. More-Than-Representational Geographies of Sound and Music.* Elgar: London. 63–74.

Anderson, J., Stoodley, L. 2017 *Beyond Water Worlds.* Royal Geographical Society Annual Conference presentation. Kensington Gore: London.

Anderson, K. 1999 Introduction. In Anderson, K., Gale, F. eds *Cultural Geographies* 2nd edn. Longman: Melbourne. 1–17.

Anderson, K., Smith, S. 2001 Editorial: Emotional geographies. *Transactions of the Institute of British Geographers* 267–10.

Anderson, T. 2007 *Riding the Magic Carpet. A Surfer's Odyssey to find the Perfect Wave.* Summersdale: Chichester.

Anderson, T. 2010 *Grey Skies, Green Waves. A Surfer's Journey around the UK and Ireland.* Summersdale: Chichester

Anderson, T. 2014 *The Actaeon Tide.* Parthian Books: Cardigan.

Anheier, H.Raj, Y. R. eds 2008 *Cultures and Globalization: The Cultural Economy.* Sage: London.

Animoto Ingersoll, K. 2016 *Waves of Knowing. A Seascape Epistemology.* Duke University Press: Durham and London.

Anzaldua, G. 1987 *Borderlands: La frontera.* Aunt Lute Books: San Francisco.

Appadurai, A. 1996 *Modernity at Large: Cultural Dimensions of Globalization.* University of Minnesota Press: Minneapolis.

Asakawa, K. 2009 Flow experience, culture, and well-being: Dow do autotelic Japanese college students feel, behave, and think in their daily lives? *Journal of Happiness Studies* 5123–154.

Aufheben, J. 1996 Review: Senseless acts of beauty. *Green Anarchist* 34http://www.geo cities.com/aufheben2/5.html.

Badiou, A. 2001 *Ethics: An Essay on the Understanding of Evil.* Verso: London.

Badiou, A. 2005 *Being and Event* Continuum: London.

Badiou, A. 2006 The Paris Commune: A political declaration on politics. In Badiou, A. ed *Polemics.* Verso: London. 257–290.

Bagshaw, G. 1998. Gapu Dhulway, Gapu Maramba: Conceptualisation and ownership of saltwater amongst the Burarra and Yan-nhangu peoples of northeast Arnhem Land. In N. Peterson, B. Rigsby eds *Customary Marine Tenure in Australia.* University of Sydney Oceania Publications: Sydney. 154–177.

Bakhtin, M. 1984 *Rabelais and his world.* Indiana University Press: Bloomington.

Bancroft, A. 1999 Gypsies to the camps!: Exclusion and marginalisation of Roma in the Czech Republic. *Sociological Research Online* 43 http://www.socresonline.org.uk/4/3/contents.html.

Barilotti S 2002 Lost horizons. Surfer colonialism in the twenty-first century. *The Surfer's Path* 3330–39.

Barilotti, S. 2013 Capturing the perfect wave: The surf photography of LeRoy Grannis. In Grannis, L. *LeRoy Grannis. Surf Photography of the 1960s and 1970s.* Taschen: Cologne. 13–19.

Barilotti, S. no date Localism works. *Surfer* http://www.surfermag.com/magazine/archivedissues/locismwrks/index3.html Accessed March 2009.

Barker, J. 2002 *Alain Badiou: A Critical Introduction*. Pluto Press: London.

Barnes, T. 2005 Culture: Economy. In Cloke, P., Johnston, R. eds, *Spaces of Geographical Thought*. Sage: London. 61–80.

Bassett, K. 2008 Thinking the event: Badiou's philosophy of the event and the example of the Paris Commune. *Environment & Planning D: Society & Space*. 26895–910.

Bauman, Z. 1995 *Life in Fragments*. Blackwell: Oxford.

Beaumont, E. 2011 *The Local Surfer: Issues of Identity and Community within South East Cornwall*. University of Exeter. Thesis for the degree of Doctor of Philosophy in Sport and Health Sciences.

Beaumont, E., Brown, D. 2016 "It's not something I'm proud of but it's … just how I feel": Local surfer perspectives of localism *Leisure Studies* 353 278–295.

Beefs TV 2019 *2 KOOKS trip out surfers at LOWERS with tricky party waves!* https://www.youtube.com/watch?v=8QJqimOa1kU Accessed December 2020.

Bennett, W. L. 1975 *The Political Mind and the Political Environment*. D. C. Heath: Lexington, MA.

Bensaid, D. 2004 Alain Badiou and the miracle of the event. In Hallward, P. ed *Think Again. Alain Badiou and the Future of Philosophy*. Continuum: London. 94–105.

Benthall, J. 2006 Archaeology and anthropology as religious movements. *Anthropology Today* 2251–2.

Blackman, L. 2012 *Immaterial Bodies: Affect, Embodiment, Mediation*. Sage: London.

Blake, T. 1935 *Hawaiian Surfriders*. Paradise of the Pacific Press: Honolulu.

Bodysurf, n.d., http://bodysurf.net/ Accessed January 2013.

Bondi, L. 1997 In whose words? On gender identities, knowledge and writing practices. *Transactions of the Institute of British Geographers* 22245–258.

Bondi, L. 2005 Making connections and thinking through emotions: between geography and psychotherapy. *Transactions of the Institute of British Geographers* 30433–448.

Booth, D. 2001 *Australian Beach Cultures: The History of Sun, Sand, and Surf*. Frank Cass: London.

Booth, D. 2001 From bikinis to boardshorts: Wahines and the paradoxes of surfing culture. *Journal of Sport History* 281 3–22.

Booth, D. 2003 Expression sessions. Surfing, style, and prestige. In Rinehart, R., Sydnor, S. eds *To the Extreme: Alternative Sports, Inside and Out*. State University of New York Press: Albany. 315–333.

Booth, D. 2004. Surfing– from one cultural extreme to another. In B. Wheaton ed *Understanding Lifestyle Sports- Consumption, Identity and Difference*. Routledge: London. 93–109.

Booth, D. 2008 Rereading the surfers' Bible: The affects of tracks. *Continuum: Journal of Media & Cultural Studies* 221 17–35.

Bottley, K. 2019 Winter wild swimming as individual and corporate spiritual practice. *Practical Theology* 123 343–344.

Bourdieu, P. 1977 *Outline of a Theory of Practice*. Cambridge University Press: Cambridge.

Bourdieu, P. 1991 *Language and Symbolic Order*. Polity Press: Cambridge.

Braun, B. 2004 Nature and culture: On the career of a false problem. In Duncan, J., Johnson, N., Schein, R. eds *A Companion to Cultural Geography*. Blackwell: Malden. 151–179.

Bredvold, R. and Skålén, P. 2016 Lifestyle entrepreneurs and their identity construction: A study of the tourism industry. *Tourism Management* 5696–105.

Britton, E. 2015 Just add surf: The power of surfing as a medium to challenge and transform gender inequalities. In Borne, G., Ponting, J. eds *Sustainable Stoke: Transitions to Sustainability in the Surfing World*. University of Plymouth Press: Plymouth. 118–127.

Brown, B. 1966 [1990 DVD] *The Endless Summer* Bruce Brown Films: Los Angeles.

Brown, B. 1997 Foreword. In Kampion, D., Brown, B. *A History of Surf Culture*. Taschen: Los Angeles 21

Brown, M., Humberstone, B. eds 2015 *Seascapes: Shaped by the sea. Embodied narratives and fluid geographies*. Ashgate: Farnham.

Brubaker, R. 1996 *Nationalism Reframed. Nationhood and the National question in the New Europe*. Cambridge University Press: Cambridge.

Buckley, R. 2002 Surf tourism and sustainable development in Indo-Pacific Islands. 1. The industry and the islands. *Journal of Sustainable Tourism* 105405–424.

Burke, E. 1776 *Philosophical Enquiry into the Origin of our Ideas of the Sublime and Beautiful*. Dodsley: London.

Buser, M., Payne, T., Edizel, O., Dudley, L. 2018 Blue space as caring space – water and the cultivation of care in social and environmental practice. *Social & Cultural Geography* 218 1039–1059.

Bush, L. 2016 Creating our own lineup: Identities and shared cultural norms of surfing women in a U.S. East Coast community. *Journal of Contemporary Ethnography* 453 290–318.

Butler, J. 1990: *Gender Trouble: Feminism and the Subversion of Identity*. Routledge: London.

Butt, T. no date Flow state, explained. *Patagonia Stories* https://eu.patagonia.com/gb/en/stories/flow/story-18147.html Accessed January 2016.

Callon, M. 1986 Some elements in a sociology of translation. In Law, J. ed *Power, Action, Belief*. Routledge: London. 19–34.

Callon, M., Law, J. 1995 Agency and the hybrid collectif. *Southern Atlantic Quarterly* 94. 483–507.

Calvino, I. 1986 *Invisible Cities*. Martin Secker and Warburg: London.

Camus, A. 2013 [1954] *The Sea Close By*. Penguin: London.

Capp, F. 2004 *That Oceanic Feeling*. Aurum: London.

Carlsen, G. A. 2019 Brazilian boys riding Indonesian waves: Orientalism and fratriarchy in a recent Brazilian surf movie. *Postcolonial Studies* 222 238–251.

Carter, P. 2018 Shadowing passage: Cultural memory as movement form. In Duxbury, N., Garret-Petts, W., Longley, A. eds *Artistic Approaches to Cultural Mapping*. Routledge: London and New York.

Casey, E. 2000 *Remembering. A phenomenological Study*. 2nd edn Indiana University Press: Bloomington.

Casey, E. 2001: Between geography and philosophy: What does it mean to be in the place-world? *Annals of the Association of American Geographers* 914 683–693.

Casey, S. 2011 *The Wave*. Anchor: New York.

Chiaroni, K. 2016 Fluid philosophy. *Performance Research* 212 108–116.

centralcoasttoday 2008 Wave Skiinghttp://www.youtube.com/watch?v=VClP4BclJic Accessed January 2021.

Clark, J. 2014 Faking it. *Drift* http://www.driftsurfing.eu/faking-it-a-closer-look-at-the-artificial-waves/ Accessed December 2014.

Cloke, P., Johnson, R. eds 2005 *Spaces of Geographical Thought: Deconstructing Human Geography's Binaries*. Sage: London.

Clough, P. 2010 The affective turn: Political economy, biomedia, and bodies. In Gregg, M., Seigworth, G. eds *The Affect Theory Reader*. Duke University Press: Durham 206–228.

Colas, A. 2001 *The World Stormrider Guide*. Low Pressure: Bude, Cornwall.

Colborn, B., Finney, B., Stallings, T., Stecyk, C., Stillman, D., Wolfe, T. 2002 *Surf Culture: The Art History of Surfing*. Gingko Press: Corte Madera, CA.

Collins Dictionary of Geology 1990 *Collins Dictionary of Geology*. Farris Lapidus, D: Oxford.

Collins English Dictionary 2003 *English Dictionary and Thesaurus*. Collins: New York.

Collins-Kreiner, N. 2010 Geographers and pilgrimages: Changing concepts in pilgrimage tourism research. *Tijdschrift voor Economische en Sociale Geografie* 1014437–448.

Comer, K. 2010. *Surfer Girls in the New World Order*. Duke University Press: Durham and London.

Comley, C. 2016 We have to establish our territory: How women surfers "carve out" gendered spaces within surfing. *Sport in Society* 198–9 1289–1298.

Conard, R. 1975 Hermann Hesse's Siddhartha, eine indische Dichtung, as a Western Archetype. *The German Quarterly* 483 358–369.

Conradson, D. 2011 The orchestration of feeling: Stillness, spirituality and places of retreat. In Bissell, D., Fuller, G. eds *Stillness in a Mobile World*. Routledge: London. 71–86.

Conradson, D. 2013 Somewhere between religion and spirituality? Places of retreat in contemporary Britain. In Hopkins, P., Kong, L., Olson, E. eds *Religion and Place: Landscape, Politics and Piety*. Dordrecht: Springer. 185–202.

Conroy, J. 2009 *Waveriders*. Inis Films and Besom Productions. Element Pictures.

Cook, J. 2004 A Voyage to the Pacific Ocean. In Warshaw, M. *Zero Break. An illustrated collection of surf writing 1777–2004*. Harvest Original / Harcourt Inc.: New York. 3–4.

Cool, N. T. 2003 *The Wetsand WaveCast ® Guide to Surf Forecasting. A Simple Approach to Planning the Perfect Sessions*. iUniverse: New York.

Cornwall Surf Guide. 2021 A to Z of surf lingo. https://www.cornwalls.co.uk/surfing/dictionary.htm Accessed May 2021.

Country, B., Wright, S., Suchet-Pearson, S., Lloyd, K., Burarrwanga, L., Ganambarr, R., Ganambarr-Stubbs, M., Ganambarr, B., Maymuru, D. 2015 Working with and learning from Country: decentring human authority. *Cultural Geographies* 222 269–283.

Crang, M., Travlou, P. 2001 The city and topologies of memory. *Environment & Planning D: Society & Space* 19161–177.

Cresswell, T. 1996 *In place/Out of place: Geography, Ideology and Transgression*. University of Minnesota Press: Minneapolis.

Cresswell, T. 2019 *Maxwell Street. Writing and Thinking Place*. University of Chicago Press: Chicago.

Croft, S. 2008 Enjoy yourself. *Drift* 22 April May 98–100.

Croker, M. Dean. J. 2012 *The Endless Winter: A Very British Surf Movie*. Level Films: London.

Crossley, N. 1996 Body-subject/body-power: Agency, inscription and control in Foucault and Merleau-Ponty. *Body and Society* 2299–116.

Csíkszentmihályi, M. 1988. The flow experience and its significance for human psychology. In M. Csíkszentmihályi, I. Csíkszentmihályi eds *Optimal experience: Psychology studies of flow in consciousness*. Cambridge University Press: Cambridge. 15–35.

Csíkszentmihályi, M. 1990. *Flow: The psychology of Optimal Experience.* Harper and Row: New York.

Csíkszentmihályi, M. 1997. *Finding Flow: The Psychology of Engagement with Everyday Life.* Harper Collins: New York.

Currie, J. 2021 Spring summer don't stop. Mark "Scratch" Cameron profile. *Wavelength* 26086–93.

Darwin, C. 1859 *On the Origin of Species.* John Murray: London.

Darwin, C. 1996 *Natural Selection.* Phoenix: London.

Daskalos, C. 2007 *Locals* only! The impact of modernity on a local surfing context *Sociological Perspectives* 501155–173.

Davidson, J. 2000 A phenomenology of fear: Merleau-Ponty and agoraphobic life-worlds. *Sociology of Health & Illness* 225 640–660.

Davidson, J. 2003 *Phobic Geographies: The Phenomenology of Spatiality and Identity.* Ashgate: Aldershot.

Davidson, J., Bondi, L., Smith, M. eds 2005. *Emotional Geographies.* Ashgate, Aldershot.

Davidson, J., Milligan, C. 2004 Editorial embodying emotion sensing space: Introducing emotional geographies. *Social and Cultural Geography* 54 523–532.

Davie, G. 2004 New approaches in the sociology of religion: A western perspective. Social Compass51173–84.

Davies, C. no date. *Surf blog.* https://catrinadavies.co.uk/category/surf/ Accessed March 2015.

Dawson, K. 2017 Surfing beyond racial and colonial imperatives in early modern Atlantic Africa and Oceania. In Zavalza Hough-Snee, D., Sotelo Eastman, A. eds *The Critical Surf Studies Reader.* Duke University Press: Durham and London. 135–154.

Delanda, M. 2006 *A New Philosophy of Society: Assemblage Theory and Social Complexity.* Continuum Books: London.

Deleuze, G. 1985 Nomad thought. In Alison, D. ed *The New Nietzsche.* MIT Press: Cambridge. 142–149.

Deleuze, G. 1993 *The Fold.* University of Minnesota Press: Minneapolis.

Deleuze, G., Guattari, F. 1981 *A Thousand Plateaus: Capitalism and Schizophrenia* University of Minnesota Press: Minneapolis, MN.

Dening, G. 2004 *Beach Crossing: Voyaging across Times, Cultures and Self.* Victoria: Carlton.

Dewsbury, J.D. 2007 Unthinking subjects: Alain Badiou and the event of thought in thinking politics. *Transactions of the Institute of British Geographers* 32443–459.

Dick-Read, A. 2007 Why?*The Surfer's Path* February/March 17.

Digance, J. 2003 Pilgrimage at contested sites. *Annals of Tourism Research* 301143–159.

Digance, J. 2006 Religious and secular pilgrimage. In Timothy, D. J., Olsen, D. H. eds *Tourism, Religion and Spiritual Journeys.* Routledge: London. 36–48.

Divine, J. 2011 *Surfing Photographs from the Eighties Taken by Jeff Divine.* T. Adler Books: Santa Barbara.

Doel, M, 1999 *Poststructuralist Geographies: The Diabolical Art of Spatial Science.* Rowman and Littlefield: Lanham, MD

Doherty, S. 2007 *The Pilgrimage: 50 Places to Surf before you Die.* Viking/Penguin: London.

Doherty, S. 2012 Mark Mathews on Fighting Fear. *Surfer* https://www.surfer.com/features/the-surfer-interview/ 27 September.

Dolnicar, S., Fluker, M. 2003 Behavioural market segments among surf tourists: Investigating past destination choice. *Journal of Sport Tourism* 84 186–196.

Dominguez Andersen, P. 2015 The Hollywood beach party genre and the exotification of youthful shite masculinity in early 1960s America. *Men and Masculinities* 185511–535.

Dovey, K. 2010 *Becoming Places. Urbanism/Architecture/Identity/Power*. Routledge: Abingdon.

Drift (no date) The Black Knight of Malibu. *Drift* https://www.driftsurfing.eu/the-black-knight-of-malibu.html Accessed December 2014.

Drift 2007 Standing tall, paddling proud. *Drift* 13 32–37.

Drift 2014 Breaking waves in the Maldives: Trouble in Paradise. *Drift* http://www.driftsurfing.eu/breaking-waves-in-the-maldives-trouble-in-paradise/ Accessed December, 2014.

Duane, D. 1996 *Caught Inside: A Surfer's Year on the Californian Coast*. North Point Press: New York.

Duggan, K. 2012 *Cliffs of Insanity*. Transworld: London.

Edwards, M. 1995 Editorial. *Wahine Magazine* 11 2–2.

eHow Sports & Fitness Editor no date How to deal with localism at surf spots. *eHow Sports & Fitness* http://www.ehow.com/how_10609_deal-with-localism.html Accessed January 2020.

Eikhof, D. R., Haunschild, A. 2006 Lifestyle meets market: Bohemian entrepreneurs in creative industries. *Creativity and Innovation Management* 153234–241.

Elden, S. 2013 *Secure the volume: Vertical geopolitics and the depth of power Political Geography*. 35–51.

Ellis, T. 2011 Agree to disagree. Knowing the feeling. *The Surfers Path* 82 130

Emerson, R. W. 1982 *Selected Essays* Penguin American Library: Harmondsworth.

Endo, T. 2013 The Surfer's Path: An Obituary *The Inertia* http://www.theinertia.com/surf/the-surfers-path-an-obituary/ December 28 Accessed December, 2013

England, S. 2014 Being Kepa *Carve* 153 August 44–55.

Entrikin, B. 1991 Hiding places. *Annals of the Association of American Geographers* 91 4694–697.

Eric 2011 Women of Bali. *The Mag* 37 May 29–32.

Esparza, D. 2016 Towards a theory of surfing expansion: The beginnings of surfing in Spain as a case study. *RICYDE. Revista Internacional de Ciencias del Deporte* XII199–215.

Evans, P. 2020 We're gonna explore the world our way. Pat O'Connell on The Endless Summer 2 *Wavelength* https://wavelengthmag.com/were-gonna-explore-the-world-our-way-pat-oconnell-on-the-endless-summer-2/ Accessed July 2020.

Evers, 2006 Locals Only!*Everyday Multiculturalism Conference Proceedings*. Macquarie University28–29 September 1–9.

Evers, C. 2006 How to Surf. *Journal of Sport and Social Issues* 30229–243.

Evers, C. 2007 Locals Only!*Everyday Multiculturalism Conference Proceedings*. Centre for Research on Social Inclusion. 1–9.

Evers, C. 2009 "The Point": surfing, geography and a sensual life of men and masculinity on the Gold Coast. *Australia Social & Cultural Geography* 10893–908.

Evers, C. 2010. *Notes for a Young Surfer*. Melbourne University Press: Carlton.

Evers, C. 2017 Surfing and contemporary China. In Hough-Snee, D. Z., Eastman, A. S. eds *The Critical Surf Studies Reader* Duke University Press: Durham. 386–402.

Farmer, B., Short, A. 2007 Australian National Surfing Reserves – rationale and process for recognising iconic surfing locations. *Journal of Coastal Research* SI 5099–103 ICS Proceedings Australia.

Featherstone, M. 1995 *Undoing Culture* London: Sage.

Feltham, O., Clemens, J. 2003 Introduction. Demonstrating Badiou. In Badiou, A. ed *Infinite Thought*. Continuum: London.

Fenster, T., Yiftachel, O. eds 1997 *Frontier Development and Indigenous Peoples*. Pergamon: Oxford.
Fiedler, L. 1966 *Love and Death in the American Novel* Stein and Day: New York.
Finnegan, W. 2015 *Barbarian Days. A Surfing Life*. Corsair: London.
Finney, B. 2001 Whoa Dude! Surfin's that Old? In Colborn, B., Finney, B., Stallings, T., Stecyk, C., Stillman, D., Wolfe, T. *Surf Culture: The Art History of Surfing*. Gingko Press: Corte Madera, CA. 82–103.
Fiske, J. 1989 *Reading the Popular*. Unwin Hyman: London.
Ford, N., Brown, D. 2006 *Surfing and Social Theory. Experience, Embodiment and Narrative of the Dream Glide*. Routledge: London.
Fordham, M. 2008 *The Book of Surfing: The Killer Guide*. Bantam Press: London.
Forsyth, I., Lorimer, H., Merriman, P., Robinson, J. 2013 Guest Editorial. *Environment and Planning A* 451013–1020.
Fox, W. 1990 *Towards a Transpersonal Ecology: Developing New Foundations for Environmentalism*. Shambala: Boston.
Franklin, R., Carpenter, L. 2019 Surfing, sponsorship and sexploitation. The reality of being a female professional surfer. In Lisahunter ed *Surfing, Sex, Genders and Sexualities*. Routledge: London and New York. 50–70.
Game, A. 1997 Sociology's emotions. *CRSA/RSCA* 344 385–399.
Game, A. and Metcalfe, A. 1996 *Passionate Sociology*. Sage: London.
Garcia, J. 2018 Orange Country. *Wavelength* 25428–33.
Gartside, L. 2019 Editorial. *Wavelength* 256 Spring Summer 18
Gartside, L. 2020 Mythology has always blossomed in communities that rely on the vagaries of the natural world. *Wavelength* Spring 2586
Germann Molz, J., Morris Paris, C. 2013 The social affordances of flashpacking: Exploring the mobility nexus of travel and communication, *Mobilities* 102 173–192.
Gibson-Graham, J.-K. 1994 Stuffed if I know! Reflections on postmodern feminist social research. *Gender, Place and Culture* 1205–224.
Glazner, E. 1995 Recollections of a world champion. *Wahine* 11 6–9.
Godlewska, A., Smith, N. eds 1994 *Geography and Empire*. Blackwell: Oxford.
Goffman, E. 1971 *Relations in Public: Microstudies of the Public Order*. Allen Lane: London.
Gosch, J. 2008 *Bustin' Down the Door* Screen Media Films: New York
Graber, L. 1976 *Wilderness as Sacred Space*. Association of American Geographers Monograph Series: Washington.
Grannis, L. 2013 *LeRoy Grannis. Surf Photography of the 1960s and 1970s*. Taschen: Cologne.
Grant, R. 2003 *Ghost Riders. Travels with American Nomads*. Abacus: London.
Green, D. C. 2007 60 paths: "How has the surfing path affected your life?" *The Surfer's Path* April/May 2007 46–61.
Greenaway, A. 2020 *Video Shows Moment 11 Surfers Ride the Same Wave*. https://www.cornwalllive.com/news/local-news/video-shows-moment-11-surfers-4233420 Accessed December 2020.
Guignon, C. C. ed 1993 *The Cambridge Companion to Heidegger*. Cambridge University Press: Cambridge.
Halbwachs, M. 1992 *On Collective Memory*. University of Chicago Press: Chicago.
Hall, S. 1996 Who needs "identity"? In Hall, S., Gay, P. eds *Questions of Cultural Identity*. London: Sage. 1–17.
Hallward, P. 2004 ed *Think Again. Alain Badiou and the Future of Philosophy*. Continuum: London.

Haraway, D. 1988 Situated knowledges: the science question in feminism and the privilege of partial perspective. *Feminist Studies* 143 575–599.

Haraway, D. 1991 *Simians, Cyborgs and Women: The Reinvention of Nature*. Routledge: New York.

Haraway, D. 2003 *The Companion Species Manifesto*. Prickly Paradigm Press: Chicago.

Haraway, D. 2016 *Manifestly Haraway: A Cyborg Manifesto*. University of Minnesota Press: ProQuest Ebook Central.

Hardin, G. 1968 The tragedy of the commons. *Science* 1623859 1243–1248.

Haro, A. 2015 Why soul surfers should love soulless wavepools. *The Inertia* http://www.theinertia.com/surf/why-soul-surfers-should-love-soulless-surf-snowdonia/#moda l-close September 27

Hayles, K. 1999. *How We Became Post-Human*. University of Chicago Press: Chicago.

Heimtun, B. 2007 Depathologizing the tourist syndrome: tourism as social capital production. *Tourist Studies* 73 271–293.

Heller, P. 2010 *Kook*. Free Press: New York.

Hemingway, J. 2006 Sexual learning and the seaside: relocating the "dirty weekend" and teenage girls' sexuality. *Sex Education* 64 429–443.

Henderson, M. 1992 What is spiritual geography? *The Geographical Review* 834469–470.

Henderson, M. 2001. A shifting line up: men, women, and tracks Surfing Magazine. *Continuum: Journal of Media and Cultural Studies* 315319–332.

Hening, G. 2015 The future of surfing is not disposable. In Borne, G., Ponting, J. eds *Sustainable Stoke: Transitions to Sustainability in the Surfing World*. University of Plymouth Press: Plymouth. 240–248.

Hetherington, K. 1998 *Expressions of Identity: Space, Performance, Politics*. Sage: London.

Hill, L. 2020/21 Unsung contributions to surf culture. *Wavelength* Autumn/Winter 25932–37.

Hill, L., Abbott, J. A. 2009 Surfacing tension: toward a political ecological critique of surfing representations. *Geography Compass* 31 275–296.

Hillier, J. 2001 Imagined value: The poetics and politics of place. In Madanipour, A., Hull, A., Healey, P. eds *The Governance of Place: Space and planning processes*. Ashgate: Aldershot. 69–101.

Hooks, B. 1991 *Yearning: Race, Gender and Cultural Politics*. Turnaround Press, New York.

Horton, D. 2003 Green distinctions: The performance of identity among environmental activists. In Szersynski, B., Heim, W., Waterton, C. eds *Nature Performed. Environment, Culture and Performance*. Blackwell: Oxford. 63–77.

Hough-Snee, D. Z., Eastman, A. S. 2017 *Consolidation, Creativity, and deColonization in the State of Modern Surfing*. In Hough-Snee, D. Z., Eastman, A. S. eds *The Critical Surf Studies Reader*. Duke University Press: Durham. 84–108.

Hough-Snee, D. Z., Eastman, A. S. eds 2017 *The Critical Surf Studies Reader*. Duke University Press. Durham.

Housman, J. 2015 Atlas be praised. *Surfer* https://www.surfer.com/features/surf-photo graphy-atlas-be-praised/ April 21 Accessed April 2015.

Hulet, S. 2011 Introduction. In Divine, J. *Surfing Photographs from the Eighties Taken by Jeff Divine*. T. Adler Books: Santa Barbara no pages.

Humberstone, B. 2015 Embodied narratives: Being with the sea. In Brown, M., Humberstone, B. eds *Seascapes: Shaped by the Sea. Embodied Narratives and Fluid Geographies*. Ashgate: Farnham27–39.

Huxley, A. 1957 *Jesting Pilate* Chatto & Windus: London.

Irons, B. 2012 The seven scariest moments of Bruce Irons' life. *Surfer* The Fear Issue August https://www.surfer.com/videos/bruceirons/ Accessed December 2012.
Jameson, F. 1991 *Postmodernism or the Cultural Logic of Late Capitalism*. Verso: London.
Jarratt, P. 2020/21 The last good season on the burkit. *Wavelength* 259 Autumn/Winter 126–135.
Jarvis, C. 2014 Every surfer should know how to: rule on a longboard. Time to embrace the many forms of surfboard. *Wavelength* January/February 23166–67.
Jarvis, C. 2014 Exclusivity is…? *Carve* 15272–77.
Jones, J. P. III, Woodward, K.Marston, S. 2007 Situating flatness. *Transactions of the Institute of British Geographers* 32264–276.
Jones, K. 2008 *Narratives of the in-between: Teenagers' identities and spatialities in a North Wales town*. Unpublished Ph.D. thesis, School of City & Regional Planning, Cardiff University. Available from the author.
Jones, M. 2009 Phase space: Geography, relational thinking, and beyond. *Progress in Human Geography* 3341–20.
Jones, P., Jones, A., Williams-Burnett, N., Ratten, V. 2017 Let's get physical: Stories of entrepreneurial activity from sports coaches/instructors. *The International Journal of Entrepreneurship and Innovation* 184219–230.
Jordan, G., Weedon, C. 1995 *Cultural Politics: Class, Gender, Race, and the Postmodern World*. Blackwell: Oxford.
Jordan, T. 2002 *Activism! Direct Action, Hacktivism and the Future of Society*. Reaktion Books: London.
Kampion, D. 2003 *The teachings of Don Redondo* http://acard.blogspot.com/2003/02/in-honor-to-all-masters-of-surfing-1.html 23 Feb 2003 . Accessed January 2013.
Kampion, D. 2004 *The Lost Coast. Stories from the Surf.* Gibbs Smith Publisher: Salt Lake City.
Kampion, D. and Brown, B. 1997 *A History of Surf Culture*. Taschen: Los Angeles.
Kobayashi, A. 2013 Critical "race" approaches. In Nuala, C., Johnson, R., Schein, H., Winders, J. eds *The Wiley-Blackwell Companion to Cultural Geography*. Oxford: Wiley Blackwell. 57–72.
Kotler, S. 2006 *West of Jesus*. Bloomsbury: London.
Kuiper, G., Smit, B. 2014 *Imagineering: Innovation in the Experience Economy*. CABI: Oxford.
Laclau, E. 2004 An ethics of militant engagement. In Hallward, P. ed *Think Again. Alain Badiou and the Future of Philosophy*. Continuum: London & New York. 120–137.
Laderman, S. 2014 *Empire in Waves*. UC Berkeley Press: Berkeley.
Lakiri Dutt, K.Samanta, G. 2013 *Dancing with the River*. Yale: New Haven.
Lambert, D., Martins, L., Ogborn, M. 2006 Currents, visions and voyages: Historical geographies of the sea. *Journal of Historical Geography* 32479–493.
Latour, B. 1993 *We Have Never Been Modern*. Harvester Wheatsheaf: Hemel Hempstead.
Latour, B. 1999 *Pandora's Hope: Essays on the Reality of Science Studies* Harvard University Press: Cambridge, MA
Latour, B. 2005 "What is given in experience?" A review of Isabelle Stenger's "Penser avec Whitehead". *Boundary* 32222–237.
Laurier, E. and Philo, C. 1999 X-morphising: review essay of Bruno Latour's Aramis, or the Love of Technology. *Environment and Planning A* 311047–1071.
Lawler, K. 2011 *The American Surfer. Radical Culture and Capitalism*. Routledge: New York.
Lazarow, N., Miller, M., Blackwell, B. 2009 The value of recreational surfing to society. *Tourism in Marine Environments* 52–3 145–158.

Leder, D. 1990 *The Absent Body*. University of Chicago Press: Chicago.
Lewis 2009 The New Localism. *Postsurf. Unfiltered Thoughts on Surf Culture*. http://postsurf.com/2009/04/01/newlocalism/ 1 April Accessed June 2009.
Lewis, N. 2000 The climbing body, nature and the experience of modernity. *Body and Society* 658–80.
Leza, J. 2012. *Diaries of a surf widow*. Accessed 18 May2017 https://www.wavescape.co.za/blog/the-surf-widow-diaries/lessons-learned.html.
lisahunter 2012 Surfing life: Surface, substructure and the commodification of the sublime. *Sport, Education and Society* 173 439–442.
lisahunter 2015 Seascapes: Surfing the sea as a pedagogy of self. In Brown, M., Humberstone, B. eds *Seascapes: Shaped by the sea. Embodied Narratives and Fluid Geographies*. Ashgate: Farnham. 41–54.
lisahunter 2019 Queering surfing from its heteronormative malaise. Public visual pedagogy of circa 2014. In lisahunter *Surfing, Sex, Genders and Sexualities*. Routledge: London and New York. 168–190.
lisahunter 2019 *Surfing, Sex, Genders and Sexualities*. Routledge: London and New York.
London, J. 1911 A royal sport, except from The Cruise of the Snark, Chapter G. http://london.sonoma.edu/Writings/CruiseOfTheSnark/snark6.html Accessed January 2012.
Lonely Planet. 2001 *South East Asia on a Shoestring*. Lonely Planet: London.
Lonely Planet. 2010 *Indonesia*. Lonely Planet: London.
Lopez, G. 2001 How to stay in the tube for the rest of your life. *The Surfers Path* 27 Oct/Nov. 26–33.
Lossau, J., Lippuner, R. 2004 Geographie and spatial turn. *Erdkunde* 58201–211.
Lovett, K. 2015 Stoke in a Sea of Uncertainty. In Borne, G., Ponting, J. eds *Sustainable Stoke: Transitions to Sustainability in the Surfing World*. University of Plymouth Press: Plymouth. 228–238.
Lukes, S. 1974 *Power: A Radical View*. Macmillan: London.
Maasen, M. 2019 Women in water: A celebration. *Wavelength* 256 Spring / Summer 24–29.
Madanipour, A., Hull, A., Healey, P.Concepts of space, concepts of place. In Madanipour, A., Hull, A., Healey, P. eds *The Governance of Place: Space and Planning Processes*. Ashgate: Aldershot. 1–21.
Magic Seaweed no date https://magicseaweed.com/.
Makarow, S. 2002 Maui simple pleasures. *Surf Life for Women* 1 1 66–73.
Malkki, L. 1992 National geographic: The rooting of peoples and the territorialization of national identity among scholars and refugees. *Cultural Anthropology* 724–44.
Malone, D. 2011 *The Secret Life of Waves*. BBC TV. https://www.youtube.com/watch?v=Kmllm1dAug4.
Maloney, W., Jordan, G., McLaughlin, A. 1994 Interest groups in public policy: The insider / outside model revisited. *Journal of Public Policy* 141 17–13.
Malthouse, J. 1999 Locals will warm up to you. *Surfing Vancouver Island* http://www.surfingvancouverisland.com/surf/st00a.htm Accessed 2009.
Mandaville, P. 1999 Territory and translocality: Discrepant idioms of political identity Millennium. *Journal of International Studies* 283653–673.
Mansfield, R. 2009 *This Surfing Tribe. A History of Surfing in Britain*. Orca: Newquay.
Mansvelt, J. 1997 Working at leisure: Critical geographies of aging. *Area* 29289–298.
Marcus, G.E. 1998 *Ethnography Through Thick and Thin*. Princeton University Press: Princeton.

Marsh, A. 2007 The art of work. *Fast Company Magazine* http://www.fastcompany.com/magazine/97/art-of-work.html Issue 97 August Accessed January 2016.

Martin, R. 1999 Critical survey: The new "geographical turn" in economics: Some critical reflections. *Cambridge Journal of Economics* 2365–91.

Massey, D. 2006 Space, time and political responsibility in the midst of global inequality. *Erkunde* 6089–95.

Massey, D. 2005 *For Space.* Sage: London.

Massey, D. 2006 Landscape as a provocation. Reflections on moving mountains. *Journal of Material Culture* 111–2:33–48.

Mattos, B. 2004. *Kayak Surfing.* Bangor: Pesda Press.

Maxam, L. 2011 Can't stop, won't stop. Kai Neville reflects on the year of Lost Atlas. *Carve* 12971–81.

Maxey, I. 1999 Beyond boundaries? Activism, academia, reflexivity and research. *Area* 31199–208.

McDowell, L. 1994 Polyphony and pedagogic authority. *Area* 263 241–248.

McDowell, L. 1999 *Gender, Place and Identity: Understanding Feminist Geographies.* University of Minnesota Press: Minneapolis.

McGloin, C. 2017 Indigenous surfing: Pedagogy, pleasure, and decolonial practice. In Zavalza Hough-Snee, D., Sotelo Eastman, A. eds *The Critical Surf Studies Reader.* Duke University Press: Durham and London.

McLuhan, T. 1996 *Cathedrals of the Spirit.* Perennial: London.

McNiven, I. 2004 Saltwater people: Spiritscapes, maritime rituals and the archaeology of Australian indigenous seascapes. *World Archaeology* 353 329–349.

Medeiros, V. 2010 Surfing is a fiction. With apologies to Jacques Lacan. *Pocmag* 195

Melucci, A. 2006 *Challenging Codes.* Cambridge University Press: Cambridge.

Melville, H. 1992 *Moby Dick.* Wordsworth: Ware

Memmott, P., Trigger, D. 1998. Marine tenure in the Wellesley Islands region, Gulf of Carpentaria. In N. Peterson, B. Rigsby eds *Customary Marine Tenure in Australia* University of Sydney Oceania Publications: Sydney. 110–124.

Merchant, S undated *Surfing and Subculture.* Unpublished thesis, University of Exeter.

Merleau-Ponty, M. 1962 *Phenomenology of Perception* Routledge & Kegan Paul: London.

Midol, N. 1993 Cultural dissents and technical innovations in the "whiz" sports. *International Review for Sociology of Sport* 28123–32.

Mihi, N. 2015 Being a brown body boarder. In Brown, M., Humberstone, B. eds *Seascapes: Shaped by the Sea.* Ashgate: Farnham. 83–100.

Minhinnick, R. 1999 *Selected Poems.* Carcanet: Manchester.

Mitchell, T. no date The seven levels of surfers. *A Spiritual and Satirical Guide Adopted to Surfing.* http://www.kenrockwell.com/tech/7surf.htm. Accessed July 2012.

Mojo, G. 2008 Shaping Mike Hynson. *Drift* 6 June/July 60–66.

Mol, A-M. 2003 *The Body Multiple.* Duke University Press: Durham, NC.

Mondy, B 2014 Harrison Roach. ~~Just another hipster cunt.~~ The world's best non-wiggly surfer? *Surf Europe* 10766–69.

Moore, T. 1992 *Care of the Soul.* Harper Perennial: New York, NY.

More, T. 1996 *The Reenchantment of Everyday Life.* Hodder and Stoughton: London.

Moriarity, J., Gallagher, C. 2001 *The Ultimate Guide to Surfing.* Lyons: London.

Murdoch, J. 2006 *Post-structuralist Geography: A Guide to Relational Space* Sage: London.

Nancy, J-L. 2000 *Being Singular Plural.* Stanford University Press: Stanford.

Nash, C. 2000 Performativity in practice: Some recent work in cultural geography. *Progress in Human Geography* 244 653–664.

Nazer, D. 2004 The tragicomedy of the surfers' commons. *Deakin Law Review* 92 655–713.

Nelson, C., Taylor, D. 2008 *Surfing Britain*. Footprint: London.

Nemani, M. 2015 Being a brown bodyboarder. In Brown, M., Humberstone, B. eds *Seascapes: Shaped by the Sea. Embodied Narratives and Fluid Geographies*. Ashgate: Farnham. 83–100.

Nijs, D., Peters, F. 2002 *Imagineering*. Boom: Amsterdam.

Nunn, T. 1998 *Tapping the Source*. Thunders' Mouth Press: New York.

Nye, J. 2004 *Soft Power: The Means to Success in World Politics*. Public Affairs: New York.

Oakes, T. 1997 Place and the paradox of modernity. *Annals of the Association of American Geographers* 87509–531.

Oelschlager, M. 1991 *The Idea of Wilderness*. New Haven, CT: Yale University Press.

Olive, R. 2015 Reframing surfing: Physical culture in online spaces. *Media International Australia*. 1551 99–107.

Olive, R. 2013 *Blurred Lines: Women, Subjectivities and Surfing*. PhD thesis, The University of Queensland.

Olive, R., McCuaig, L., Phillips, M. 2015 Women's recreational surfing: A patronising experience. *Sport, Education and Society* 202 258–276.

Olive, R., Roy, G., Wheaton, B. Stories of surfing. Surfing, space, and subjectivity/intersectionality. In Lisahunter *Surfing, Sex, Genders and Sexualities*. Routledge: London and New York. 148–168.

Orams, M., Towner, N. 2013 Riding the wave: History, definitions, and a proposed typology of surf-riding tourism. *Tourism in Marine Environments* 84 173–188.

Ormrod, J. 2005 Endless summer 1964: Consuming waves and surfing the frontier. *Film & History: An Interdisciplinary Journal of Film and Television Studies* 351 39–51.

Ormrod, J. 2007 Surf rhetoric in American and British surfing magazines between 1965 and 1976. *Sport in History* 271 March 88–109.

Otto, R. 1970 *Mysticism East and West: A Comparative Analysis of the Nature of Mysticism*. Macmillan: New York, NY.

Ouhilal, Y. 2011 Utopia. A place that cannot be. *Carve* 123 May 88–97.

Oxford Dictionary 2011 *English Dictionary*. Oxford University Press: Oxford.

Page, G. 2003 Vintage days in the big waves of life. In Rinehart, R., Sydnor, S. eds *To the Extreme: Alternative Sports, Inside and Out*. State University of New York Press: Albany. 307–314.

Pantzar, M., Shove, E. 2004 *Manufacturing Leisure. Innovations in Happiness, Well-being and Fun*. National Consumer Research Centre: Helsinki.

Pearson, K. 1982. Conflict, stereotypes and masculinity in Australian and New Zealand surfing. *Australian and New Zealand Journal of Sociology* 182 117–135.

Peralta, S. 2004 *Riding Giants*. BoB – On Demand TV and Radio for Education Learning on Screen – The British Universities and Colleges Film and Video Council My Bob Guide. https://learningonscreen.ac.uk/bob/

Peters, K. 2012 Manipulating material hydro-worlds: Rethinking human and more-than human relationality through offshore radio piracy. *Environment and Planning A* 441241–1254.

Peters, K. 2010 Future promises for contemporary social and cultural geographies of the sea. *Geography Compass* 49 1260–1272.

Peters, K., Steinberg, P. 2015 A wet world: Rethinking place, territory, and time. *Society & Space* http://societyandspace.org/2015/04/27/a-wet-worldrethinking-place-territory-and-time-kimberley-peters-and-philip-steinberg Accessed 8 August2018.

Peters, K., Steinberg, P. 2019 The ocean in excess: Towards a more than wet ontology. *Dialogues in Human Geography* 93https://doi.org/10.1177/2043820619872886.

Peters, K., Steinberg P., Stratford, E. eds 2018 *Territory Beyond Terra*. Rowman and Littlefield: London.

Pettibon, R. 1989 *It is in the waves that he states his ideas most unmistakably*. Serigraph, 22–21/2x17inches. Collection of Laguna Art Museum, Orange County Museum of Art.

Phillips, D. 2006 Parallel lives? Challenging discourses of British Muslim self-segregation. *Environment & Planning D, Society & Space* 2425–40.

Philo, C. 2005 Editorial. Spacing lives and lively spaces: Partial remarks on Sarah Whatmore's Hybrid Geographies. *Antipode* 824–833.

Pile, S. 1994. Masculinism: The use of dualistic epistemologies and third spaces. *Antipode* 263 5–43.

Pill, E. 2019. *Waves of power. The spectacularisation of professional surfing*. PhD Thesis, Cardiff University.

Point Break 1991 *Transcript*. http://www.script-o-rama.com/movie_scripts/p/point-break-script-transcriptkeanu.html. Director: Bigelow, J. Accessed June 2009.

Ponting J. 2008 Consuming Nirvana: An exploration of surfing tourist space (doctoral dissertation). University of Technology, Sydney.

Ponting, J. 2009 Projecting Paradise: The surf media and the hermeneutic circle in surfing tourism. *Tourism Analysis* 14175–185.

Ponting, J., McDonald, M., Wearing, S. 2005 Deconstructing wonderland: Surfing tourism in the Mentawai Islands, Indonesia. *Society & Leisure* 281 141–162.

Ponting, J., O'Brien, D. 2014 Liberalizing Nirvana: An analysis of the consequences of common pool resource deregulation for the sustainability of Fiji's surf tourism industry. *Journal of Sustainable Tourism* 223384–402.

Pratt, G. 2004 Feminist geographies: Spatialising feminist politics. In Cloke, P., Crang, P., Goodwin, M. eds *Envisioning Human Geographies*. Hodder Arnold: London. 283–304.

Preston, C. 2003 *Grounding Knowledge: Environmental Philosophy, Epistemology, and Place*. University of Georgia Press: Athens and London.

Preston-Whyte, R. 2001. Constructed leisure Space: The seaside at Durban. *Annals of Tourism Research* 283581–596.

Preston-Whyte, R. 2002 Constructions of surfing space at Durban, South Africa. *Tourism Geographies* 4307–328.

Preston-Whyte, R. 2008 The beach as liminal space. In Lew, C., Hallam, C. M., Williams, A. eds *A Companion to Tourism*. Blackwell: Oxford. 349–359.

Pretor-Pinney, G. 2010 *The Wave-watcher's Companion*. Perigee Books: New York.

Price, J. 1996 Naturalistic recreations. In H. Van Ness, P. ed *Spirituality and the Secular Quest*. Crossroad: New York, NY. 414–444.

Pugh, J. 2009 Viewpoint: what are the consequences of the "spatial turn" for how we understand politics today? A proposed research agenda. *Progress in Human Geography* 33579–586.

Pumphrey, N. 2014 Meet Mike Lay. *Carve* 152 July 38–43.

Raban, J. 1999 *Passage to Juneau. A Sea and its Meanings*. Picador: London.

Radley, A. 1995 The elusory body and social constructionist theory. *Body and Society* 123–23.

Relph, E. 1976 *Place and Placelessness*. Pion: London.

Remy, R. 1990 Patriarchy and fratriarchy as forms of androcracy. In Hearn, J., Morgan, D. eds *Men, Masculinities and Social Theory*. Unwin Hyman: London.

Revill, G. 2016. How is space made in sound? Spatial mediation, critical phenomenology and the political agency of sound. *Progress in Human Geography* 402 240–256.

Rich, L. 2006 Rival surf gangs fight for the waves. *The First Post* http://www.thefirstpost.co.uk/2726,news-comment,news-politics,rival-surf-gangs-fight-for-the-waves. August 10. Accessed January 2008.

Richards, M. 2002 Real surfing. Agree to disagree. *The Surfer's Path* 28 December-January 114

Rickly-Boyd, J. 2012 Lifestyle climbing: toward existential authenticity. *Journal of Sport and tourism* 17285–104.

Rinehart, R. 2015 Surf film, then and now: The Endless Summer meets Slow Dance. *Journal of Sport and Social Issue* 396545–561.

Rinehart, R., Sydnor, S. eds 2003 *To the Extreme: Alternative Sports, Inside and Out*. State University of New York Press: Albany, New York.

Roberts, M., Ponting, J. 2018 Waves of simulation: Arguing authenticity in an era of surfing the hyperreal. *International Review for the Sociology of Sport* 1–17.

Robertson, D. 2019 Why feminism in surf is needed – A response. [Weblog] Cardiff University Blog: Surfing Research. 30/20/2019. Available from: https://blogs.cardiff.ac.uk/surfing-research/why-feminism-in-surf-is-needed-a-response.

Robertson, K. 2020 Surfer Magazine's long ride may be over. *The New York Times* Oct 5. https://www.nytimes.com/2020/10/05/business/media/surfer-magazine.html Accessed October 2020.

Robins, R. P., Stock, E. C., Trigger, D. S. 1998. Saltwater people, saltwater country: Geomorphological, anthropological and archaeological investigations of the coastal lands in the southern Gulf of Carpentaria country of Queensland. *Memoirs of the Queensland Museum Cultural Heritage Series* 1175–125.

Rode, M. 2020 Is sharing caring when it comes to XXL waves? *Magic Seaweed* https://magicseaweed.com/news/when-is-it-ok-to-share-a-xxl-wave/12208/ 28 December 2020.

Rogatko, T. 2009 The influence of flow on positive affect in college students. *Journal of Happiness Studies* 10133–148.

Rose, G. 1997 Situating knowledges: Positionality, reflexivities and other tactics. *Progress in Human Geography* 21305–320.

Rouse, J. 1996 *Engaging Science: How to Understand its Practices Philosophically* Cornell University Press: Ithaca, NY.

Routledge, P. 1996 Third space as critical engagement. *Antipode* 284 399–419.

Routledge, P. 1997 The Imagineering of resistance. *Transactions of the Institute of British Geographers*. 223 359–376.

Routledge, P., Cumbers, A., Nativel, C. 2007 Grassrooting network imaginaries: Relationality, power, and mutual solidarity in global justice networks. *Environment and Planning A* 39 11 2575–2592.

Ruttenberg, T., Brosius, P. 2017 Decolonising Sustainable Surf Tourism. In Zavalza Hough-Snee, D., Sotelo Eastman, A. eds *The Critical Surf Studies Reader*. Duke University Press: Durham and London. 109–132.

Ryan, A. 2012 *Where Land Meets the Sea*. Routledge: London and New York.

Sack, R. D. 1997 *Homo Geographicus: A Framework for Action, Awareness and Moral Concern*. Johns Hopkins University Press: Baltimore.

Said, E. 1993 *Culture and Imperialism*. Knopf: New York.

Sanders, M. no date. Localism: Not just for idiots anymore? *Groundswell Society* http://www.groundswellsociety.org/events/SASIC/pdf/SASIC%201%20-%20Localism%20by%20Marcus%20Sanders.pdf Accessed June 2009.

Sanford, A.W. 2007 Pinned on Karma rock: Whitewater Kayaking as religious experience. *Journal of the American Academy of Religion* 754875–895.

Sargisson, L. 2000 *Utopian Bodies and the Politics of Transgression.* Routledge: London.

Sassen, S. 1996 *Losing control: Sovereignty in an Age of Flobalization.* Columbia University Press, New York.

Scales, J. 2015 Science goes in search of the perfect wave BBC News. http://www.bbc.co.uk/news/science-environment-34054322 9 September 2015.

Scheibel, D. 1995 Making waves with Burke: Surf nazi culture and the rhetoric of localism. *Western Journal of Communication* 594 253–269.

Seamon, D. 1979 *A Geography of the Lifeworld: Movement, Rest and Encounter.* Routledge: London and New York.

Seamon, D., Mugerauer, R. eds 1985 *Dwelling, Place and Environment: Toward a Phenomenology of Person and World.* Kluwer: Malabar, Florida.

Searle, L. 2015 Contents. *Surfgirl* 5214.

Seifert, T., Hedderson, C. 2009 Intrinsic motivation and flow in skateboarding: An ethnographic study. *Journal of Happiness Studies* doi:10.1007/s10902-009-9140-y110–125.

Severson, J. 2014 *John Severson's Surf.* Puka: Hawai'i.

Sharp, N. 2002. *Saltwater People.* Allen & Unwin: Crows Nest.

Sharp, R. 2014 The life cycle of a wave. *Carve* 153 August 56–63.

Sharp, R. 2015 Editorial. *Carve* 15711.

Sheller, M., Urry, J. 2004 Places to play, places in play. In Sheller, M. and Urry, J. eds *Tourism Mobilities: Places to Play, Places in Play.* Routledge: London. 1–10.

Sheller, M., Urry, J. 2006 The new mobilities paradigm. *Environment and Planning A* 38207–226.

Shields, R. 1991 *Places on the Margin: Alternative Geographies of Modernity.* Routledge: London.

Simmel, G. 1950. *The Sociology of Georg Simmel.* Free Press: London.

Slater, E. 2008 Foreward. *Surfing Magazine* http://www.surfingmagazine.com/magazine/july-issue-2008-surfingmagazine/ Accessed July 2012.

Smith, C. 2013 *Welcome to Paradise: Now Go to Hell.* Dey Street Publishers: New York.

Smith, J. 1998 Religion, religions, religious, In Taylor, C. ed *Critical Terms for Religious Studies.* University of Chicago Press: Chicago, IL. 269–284.

Smith, J., Green, W. 1995 Religion, definition of. In Smith, J. Z. and Green, W. S. eds *The HarperCollins Dictionary of Religion.* HarperCollins: New York, NY. 893–894.

SoCal Surfer 2019 Surfers Catch INSANE Party waves in San Diego!https://www.youtube.com/watch?v=EaZ-uNoMRHw Accessed December 2020.

Soja, E. 1996 *Thirdspace: Journeys to Los Angeles and Other Real-and-Imagined-Places.* Blackwell: Maldon.

Soja, E. 2010 *Seeking Spatial Justice.* University of Minnesota Press: Minneapolis.

Sotomayor, S., Barbieri, C. 2016 An exploratory examination of serious surfers: Implications for the surf tourism industry. *International Journal of Tourism Research* 181 62–73.

Stannard, D. 1989 *Before the Horror: The Population of Hawai'i on the Eve of Western Contact.* University of Hawaii Press: Honolulu.

States Symbols USA no date *Official State Aloha Spirit of Hawai'I* https://statesymbolsusa.org/symbol-official-item/hawaii/cultural-heritage/aloha-spirit#:~:text=%22Aloha%22%20means%20mutual%20regard%20and,and%20to%20know%20the%20unknowable.%22.

Stecyk, C. 2002 Introduction. In Colborn, B., Finney, B., Stallings, T., Stecyk, C., Stillman, D., Wolfe, T. *Surf Culture: The Art History of Surfing.* Gingko Press: Corte Madera, CA. 32–81.

Stedman, L. 1997 From gidget to gonad man: Surfers, feminists and postmodernisation. *ANZJS* 33 182–83.

Steinberg, P. 2001 *The Social Construction of the Ocean*. Cambridge University Press: Cambridge.

Steinberg, P. 2013 Of other seas: Metaphors and materialities in maritime regions. *Atlantic Studies* 10 156–169.

Steinberg, P. 2015 Foreword. In Brown, M., Humberstone, B. eds *Seascapes: Shaped by the sea. Embodied narratives and fluid geographies*. Ashgate: Farnham. xi–xiv.

Steinberg, P., Peters, K. 2015 Wet ontologies, fluid spaces: Giving depth to volume through oceanic thinking. *Environment and Planning D: Society and Space* 33 2247–264.

Sterling, B. 2020 Romanticism in surf mythology. *Wavelength* Spring 258 36–41.

Storr, W. 2019 *The Science of Storytelling*. William Collins: London.

Stranger, M. 1999 The aesthetics of risk: a study of surfing *International Review for the Sociology of Sport* 34 3265–276.

Stranger, M. 2011 *This Surfing Life. Surface Substructure and the Commodification of the Sublime*. Ashgate: Farnham.

Stronach, I., Maclure, M. 1997 *Educational Research Undone: The Postmodern Condition*. Milton Keynes: Open University.

Surfer (no date) Cover Archive. *Surfer* https://www.surfer.com/cover-archive/ Accessed January 2014.

Surfer Today 2020 Surfer Magazine closes after 60 years. *Surfer Today* https://www.surfertoday.com/surfing/surfer-magazine-closes-after-60-years Accessed January 2021.

Surfer Today no date The surfing magazines of the world. *Surfer Today* https://www.surfertoday.com/surfing/the-surf-magazines-of-the-world Accessed January 2021.

Surfer Today. 2013 The forces of power and influence in surfing. *Surfer Today* http://www.surfertoday.com/surfing/9332-the-forces-of-power-and-influence-in-surfing 30 September Accessed January 2015.

Surfer Today 2021 The basic rules of surf etiquette. *Surfer Today* https://www.surfertoday.com/surfing/the-basic-rules-of-surf-etiquette Accessed January 2021.

Surfer Today no date How does wave priority work in surfing? *Surfer* https://www.surfertoday.com/surfing/how-does-wave-priority-work-in-surfing. Accessed January 2020.

Surfer Today no date The best surfing quotes of all time. *Surfer Today* https://www.surfertoday.com/surfing/the-best-surfing-quotes-of-all-time Access January 2013.

Surferpedia no date The Endless Summer. *Surferpedia* https://surfing.fandom.com/wiki/The_Endless_Summer Accessed January 2019.

Surflife for Women 2002 Surflife for women. *Premier Issue* https://issuu.com/historyofwomensurfing/docs/premierissue Accessed January 2021.

Surfline, no date. Bra boys on localism. *Surfline* http://www.surfline.com/video/webisodes/bra-boys-on-localism_11560 Accessed June 2009.

Taylor, B. 2007a Focus introduction. Aquatic nature religion. *Journal of the American Academy of Religion* 75 4863–874.

Taylor, B. 2007b Surfing into spirituality and a new, aquatic nature religion. *Journal of the American Academy of Religion* 75 4923–951.

Taylor, B. 2007c The new aquatic nature religion. *Drift* 1 314–23.

Taylor, B. 2008 Sea spirituality, surfing, and aquatic nature religion. In Shaw, S., Francis, A. eds *Deep Blue: Critical Reflections on Nature, Religion and Water*. Equinox: London. 213–233.

Taylor, K. 2005 *Return by Water. Surf Stories and Adventures*. Dimdim Publishing: Diego.

Taylor, K. 2019 Lie down and be counted. *Wavelength* 256 Spring Summer 94–107.

Taylor, K. no date *The surf industry stole my culture* https://kimballtaylor.bigcartel.com/product/the-surf-industry-stole-my-culture-and-all-i-got-was-this-stupid-hat.

The Economist. 2012 Beach rush. http://www.economist.com/node/21550253 Accessed January 2013.

Thomas, B. 2014 Editor: A little privacy. *Surfer* 55 20

Thomson, G. 2017 Pushing under the whitewash: Revisiting the making of South Africa's surfing sixties. In Hough-Snee, D. Z., Eastman, A. A. eds *The Critical Surf Studies Reader*. Duke University Press: Durham and London. 155–176.

Thoreau, H. D. 2018 *Walden and Civil Disobedience*. William Collins: London.

Thorne, T. 1976a Legends of the surfer subculture: Part one. *Western Folklore* 353 July 209–217.

Thorne, T. 1976b Legends of the surfer subculture: Part two. *Western Folklore* 354 October 270–280.

ThreeSixty, no date http://www.threesixtymag.co.uk/ Accessed January 2013.

Thrift, N. 1996 *Spatial Formations*. Sage: London.

Thrift, N. 1997 The still point: Resistance, expressive embodiment and dance. In: Pile, S., Keith, M. eds *Geographies of Resistance*. Routledge: London. 124–151.

Thrift, N. 1999 Steps to an ecology of place. In Massey, D., Allen, J., Sarre, P. eds *Human Geography Today*. Polity Press: Cambridge. 295–322.

Tilley, C. 1994 *A Phenomenology of Landscape: Places, Paths and Monuments*. Berg: Oxford.

Tomlinson, J. 2001 *Extreme Sports: The Illustrated Guide to Maximum Adrenaline Thrills*. Carlton Books: London.

Tuan, Y.-F. 1975 *Topophilia: A Study of Environmental Perception, Attitudes, and Values*. Prentice Hall: Englewood Cliffs.

Turner, V. 1982 Liminal to liminoid in play, flow, ritual: An essay in comparative symbology. In Turner, V. ed *From Ritual to Theatre: The Human Seriousness of Play*. Performing Arts Journals Publications: New York.

Twain, M. 2004 Roughing it. In Warshaw, M. *Zero Break. An Illustrated Collection of Surf Writing 1777–2004*. Harvest Original / Harcourt Inc.: New York. 6–8.

United Nations 2017 *Progress towards the Sustainable Development Goals Report of the Secretary-General*. United Nations Economic and Social Research Council https://unstats.un.org/sdgs/files/report/2017/secretary-general-sdg-report-2017-Statistical-Annex.pdf.

US Department of State 2001–2009 Archive. https://2001-2009.state.gov/r/pa/ho/time/gp/17661.htm#:~:text=Spurred%20by%20the%20nationalism%20aroused,Dole%20became%20its%20first%20governor.

van Gennep, A. 1909 [1960] *The Rites of Passage: A Classic Study of Cultural Celebration*. The University of Chicago Press: Chicago.

Wade, A. 2007 *Surf Nation*. Simon & Schuster: London.

Waitt, G. 2008 "Killing waves": Surfing, space and gender. *Social & Cultural Geography* 9:1, 75–94.

Waitt, G., Clifton, D. 2013 "Stand out, not up": Bodyboarders, gendered hierarchies and negotiating the dynamics of pride/shame. *Leisure Studies* 325 487–506.

Waitt, G., Frazer, R. 2012 The vibe and the glide: Surfing through the voices of longboarders. *Journal of Australian Studies* 363 327–343.

Walker, I. 2011 *Waves of Resistance: Surfing and History in Twentieth-Century Hawai'i* University of Hawai'i Press: Honolulu.

Wallace, J. 2003 A Karl, not Groucho Marxist in Springfield. In Irwin, W., Conard, M., Skoble, A. eds *The Simpsons and Philosophy: The D'oh of Homer*. Open Court: Chicago. 235–251.

Wallis, L., Walmsley, A., Beaumont, E., Sutton., C. 2020 "Just want to surf, make boards and party": How do we identify lifestyle entrepreneurs within the lifestyle sports industry. *International Entrepreneurship and Management Journal* https://doi.org/10.1007/s11365-020-00653-22020.

Ward, N. 1996 Surfers, sewage, and the new politics of pollution. *Area* 283 331–338.

Warf, B. and Arias, S. eds 2008 *The Spatial Turn: Interdisciplinary Perspectives* Routledge: London.

Warren, A., Gibson, C. 2014 *Surfing places, Surfboard makers. Craft, Creativity, and Cultural Heritage in Hawai'i, California, and Australia.* University of Hawai'i Press: Honolulu.

Warshaw, M. 2004 *Zero Break. An illustrated collection of surf writing 1777–2004.* Harvest Original / Harcourt Inc: New York.

Warshaw, M. 2005 *The encyclopedia of surfing.* Harvest: Orlando.

Warshaw, M. 2010 *The History of Surfing.* Chronicle: San Francisco.

Watson, J. W. 1983 The soul of geography. *Transactions of the Institute of British Geographers* 8385–399.

Watson, M. Shove, E. 2008 Product, competence, project and practice: DIY and the dynamics of craft consumption. *Journal of Consumer Culture* 869–89.

Watson, Sean. 1998 The neurobiology of sorcery: Deleuze and Guattari's brain. *Body and Society* 4423–45.

Watters, R. 2003. The wrong side of the thin edge. In Rinehart, R., Sydnor, S. eds *To the extreme: Alternative sports inside and out.* State University of New York Press: Albany, New York. 257–266.

Weisbecker, A. 2001 *The Search for Captain Zero.* Tarcher: Los Angeles.

Weisberg, Z. 2020 Surfer Magazine just published its last issue. *The Inertia* 3rd October https://www.theinertia.com/surf/surfer-magazine-just-published-its-last-issue/ Accessed November 2021.

West, P. 2014 Such a site for play, this edge: Surfing, tourism, and modernist fantasy in Papua New Guinea. *The Contemporary Pacific* 262 411–432.

Westwick, P., Neushel, P. 2013 *The World in the Curl. The Unconventional History of Surfing.* Crown: New York.

Wheaton, B. 2004 Introduction: Mapping the lifestyle sportscape. In Wheaton, B. ed *Understanding Lifestyle Sports: Consumption Identity and Difference.* Routledge: London1–28.

Wheaton, B. 2007 Identity, politics, and the beach: Environmental activism in Surfers against Sewage. *Leisure Studies* 263 279–302.

Wheaton, B. 2017 Space invaders in surfing's white tribe: Exploring surfing, race, and identity. In Zavalza Hough-Snee, D., Sotelo Eastman, A. eds *The Critical Surf Studies Reader.* Duke University Press: Durham and London. 177–196.

Wheaton, B. 2018 The cultural politics of surfing spaces: stories of "local" identity and belonging in Aotearoa/New Zealand. Surfing Beyond Cliches: New Water Cultures Symposium Cardiff University.

Whitman, W. 2003 *Leaves of Grass.* Dover Publications: New York.

Wilson, B. 2008 *Dances with Waves.* Two Ravens Press: Ullapool.

Wilson, S. 2017 The composition of posthuman bodies. *International Journal of Performance Arts and Digital Media* 132 137–152.

Winton, T. 1993 *The Land's Edge.* Picador: London.

Winton, T. 2008 *Breath.* Picador: London.

Wolfe, T. 2002 The Pump House Gang. In Colborn, B.Finney, B.Stallings, T.Stecyk, C.Stillman, D.Wolfe, T. *Surf Culture: The Art History of Surfing.* Gingko Press: Corte Madera, CA. 104–115.

Wood, N., Smith, S. 2004 Instrumental routes to emotional geographies. *Social and Cultural Geography* 54 533–548.

Woody, T. 2015 It's time to bring the image of the surfer into the 21st century. In Borne, G., Ponting, J. eds *Sustainable Stoke. Transitions to Sustainability in the Surfing World* University of Plymouth Press: Plymouth. 70–73.

Yarwood, R.Shaw, J. 2010 "N-gauging" geographies: Craft consumption, indoor leisure and model railways. *Area* 424 425–433.

Yogis, J. 2009 Everything is a wave. *The Surfers Path*. http://surferspath.mpora.com/features/agree-to-disagree/everything-is-a-wave.html 20 March

Young, N. 2007 60 Paths: How has the surfing path affected your life? *The Path* April / May 46–61.

Zavalza Hough-Snee, D. and Sotelo Eastman, A. 2017 Introduction. In Zavalza Hough-Snee, D., Sotelo Eastman, A. eds *The Critical Surf Studies Reader*. Duke University Press: Durham and London. 1–28.

Zhou, N. 2020 "Remarkable": South Australian surfer with serious shark bite injuries swims to shore and walks 300m. *The Guardian* https://www.theguardian.com/environment/2020/dec/07/remarkable-south-australian-surfer-with-serious-shark-bite-injuries-swims-to-shore-and-walks-300m 7 December. Access December 2020

Zink, R. 2015 Sailing across the Cook Strait. In Brown, M., Humberstone, B. eds *Seascapes: Shaped by the Sea. Embodied Narratives and Fluid Geographies*. Ashgate: Farnham 71–82.

Zizek, S. 2004 From purification to subtraction: Badiou and the real. In Hallward, P. ed *Think Again. Alain Badiou and the Future of Philosophy*. Continuum: London and New York. 165–181.

Index

Aguerre 71, 152
Alexander 203
Allen 3, 33, 48,
Anderson, A. 123, 129, 133
Anderson, B. 47, 169,
Anderson, J. 5, 6, 16, 75, 81, 84, 105, 107, 118, 152, 159, 196
Anderson, T. 87, 93, 106, 107, 165, 170, 196
aquatic nature religion 75, 86,
architecture (media) 19–21, 126, 137, 139–147, 188–191
assemblage 19–20, 29
attentiveness 75–77

Badiou 102–104
Barilotti 147, 165, 176, 196–200
Bauman 164
beach 40
Beaumont 168
Blake 122, 127
bodyboarding 65–66, 174, 180
bodysurfing 64
Bondi 92
Booth 151, 153, 180, 184,–186
Bourdieu 178
Britton 189, 192, 204
Brody 204
Brown, B. 14, 70, 84, 90–91, 128–129, 134, 141, 143–145
brown girl surf 204
Brubaker 102
Burkard 213
Bush 186, 188, 191, 211

californication of surfing 92
Cameron 169
Capp 9, 16, 49, 57, 82–84
Carlsen 198, 201

carnivalesque 15, 41
Casey, E. 53–54
Casey, S. 43, 55, 91
climbing body 60
codes 20, 22
colonial surf-riding 118
colonisation through conquest 118–120
colonisation through cultural creativity 122–123
colonisation through cultural entrepreneurship 120–122
Comer 127, 132, 134–135
Comley 186
Conard 136
convergence 19, 60, 75, 81–83
Cook 94, 197
craft 11, 21, 66, 117, 171–181
Croft 217
Csíkszentmihályi 77–80
cultural geography of surfing spaces 4–6
culturalism 115, 201
cyborg 19, 63–72

Dau Hui 166
Dawson 116, 121
definition (of surf-riding) 11
Dick-Read 85
disorientation 81
Divine 132, 182, 212–213, 217
Dora 218
Duane 15–16, 55, 93
Duggan 15–16

eastern archetype 192
Edwards, P. 7, 91, 141, 144
Elden 48, 174
embodied turn 26–28
entrepreneurship (surf-lifestyle) 20, 71, 120, 125–137

Entrikin 52
escapism 15
event 102–109
Evers 57–58, 81, 90, 98, 151, 164, 175, 185

fast food surfing 216
fear 13, 90, 95, 97–99
fidelity 103–109
Fiedler 136
Finnegan 117–118, 140, 161
flat ontology 48
flow 77–80
Ford and Brown 60, 84, 90, 92, 156
Fox 59
fratriarchies 185, 217

Gamson and Wolfsfeld 127, 136, 140, 143
gender 176, 182–193
Grannis 184
Grimes 193

Haraway 19, 30, 63, 68
Haro 214–215
Hayles 63
Helekunihi Walker 119–120
Heteronormativity 131, 176
Hill 183, 188
histories of surf-riding 113–124
homophobia 175, 181
Horton 126, 147–148, 151
Hough-Snee and Eastman 23, 114, 146–7, 188, 204
Hume Ford 95, 120–123
Humberstone 56
Huxley 137
hydro turn 25–26
hydro-humans 55–56

imagineering 125–138
indigeneity and identification with water 57–61
Ingersoll 58–59, 61

Jarvis 179, 219
Johnson 68
Jordan and Wheedon 32, 208

Kampion 33, 57, 69, 70, 93, 123, 128, 129,
Kampion and Brown 14, 16, 70, 90, 91, 141, 143
Kingsley 166
Kotler 46, 82, 83, 107

Laderman 21, 121, 135, 218
Latour 27, 29
Lawler 122, 142, 144
Lay 172
Lewis 60
liminality 41
Lisahunter 60, 176
littorality 4, 14, 40–43
littorality as zone of energy transfer 44–45
location of waves 39–51, 214–217
London 65, 95–97
longboards 69–70
Lopez 177

male gaze 185
Malthouse 168
Mansfield 6, 7, 15, 141, 142
McGloin 204
Melucci 148, 151
Melville 22
Merchant 187
Merleau-Ponty 27
metropolitan body 60
military metaphors 177–178
Minhinnick 95
Mixophobia 164–165

Noyle 212

Olive 33, 183, 186, 213
Olive, McCuaig, and Phillips 187, 203
one surfer, one wave 155
Ormrod 129, 132, 138
Outback 96

Page 97, 105
paradox of representation 91
patriarchy 175, 186
Pearson 184
Pepin Silva 128, 152, 185
perfect ride 99
performance capital 177
pilgrimage 87
politics of verticality 173–174
Ponting 6, 10, 157, 197, 204, 209
positionality 18, 21, 25, 32–36, 150–159
pre-colonial surf-riding 113–115
Preston-Whyte 40, 42
prosthesis 67
provenance 21, 160–169

racism 201, 203
relational sensibility 75, 76, 81, 84
relational turn 28

relationality 51
religion 19, 75, 81, 85, 87
riding technology 11, 52, 72
Rinehart 184–185
risks 8, 9, 97
Robertson 140, 192
Routledge et al 127
Ruttenberg and Brosius 201

Scales 215
Scheibel 81, 156, 162
script 19–22, 125–149
script and codes 150–218
Seamon 76–77
seascape epistemology 60
sedentary metaphysics 26
Severson 71, 128–137
shortboard 71, 176, 178, 180, 215
Simmel 163
sit-on-top kayaks 11, 66
Smith 16, 133, 147, 203
snaking 156–157
Soja 24, 34
source 84
spatial turn 24, 25
spectacle of surf–riding 94
spirituality 85–87
stand up paddle boards 155, 174, 179
Stecyk 116, 117, 119
Stedman 148, 184
stem script 208–218
stoke 19, 75, 89–101
Storr 190
Stranger 9, 78, 83, 143, 146, 193, 213,
sublime 6, 83–84
surf kayaking 66–67
surf magazines and gender 177, 184, 188–191
surf skis 66
surf studies 23

Surfer magazine 71, 130, 140, 142, 149
surf-riding as compulsion 107
surf-riding body 60

Taylor, B 75, 84–87
Taylor, K. 173–175, 179–181
terracentric human geography 40
The Black Surfers' Collective 204
The Endless Summer 13, 105, 134–138, 142, 196, 198
thirdspace 34–36
Thomson, S 79
Thorne 9, 131, 151
Thresholders 146
trans–local surf–rider 195–205
Trent, B. 186, 187
Twain 10, 94

violence 156–157, 164–167

Wade 10, 86, 97, 157, 165, 166
Wahine 188–190
Waitt & Clifton 65, 175, 180
Warren and Gibson 11, 72, 120, 177
Warshaw viii, 19, 92, 93, 115, 116, 117, 122, 123, 146
water people 55–56
wave sharing 154, 157, 168
waves (formation of) 39–50
Weizbecker 99
Weizman 174
West 198
western archetype 135, 185, 188, 197
western frontier 132–133
Wheaton 59, 202
Whitman 3
wild surf-riding 17, 55, 216
Winton 4, 13, 14, 57, 101
Wolfe 140, 143, 144
Woody 173, 209, 210

For Product Safety Concerns and Information please contact our EU representative GPSR@taylorandfrancis.com Taylor & Francis Verlag GmbH, Kaufingerstraße 24, 80331 München, Germany

Printed and bound by CPI Group (UK) Ltd, Croydon, CR0 4YY
08/06/2025
01897009-0010